Plym
Charles
Subject to stat
via y

http://prim
Tel: (01

Ocean Engineering & Oceanography

Volume 6

Series editors

Manhar R. Dhanak, Florida Atlantic University SeaTech, Dania Beach, USA
Nikolas I. Xiros, New Orleans, USA

More information about this series at http://www.springer.com/series/10524

Benoît Zerr · Luc Jaulin · Vincent Creuze
Nathalie Debese · Isabelle Quidu
Benoît Clement · Annick Billon-Coat
Editors

Quantitative Monitoring of the Underwater Environment

Results of the International Marine Science
and Technology Event MOQESM'14
in Brest, France

Springer

Editors
Benoît Zerr
Ocean Sensing and Mapping
ENSTA Bretagne
Brest Cedex 09
France

Luc Jaulin
ENSTA Bretagne
Brest Cedex 09
France

Vincent Creuze
Robotic Department
University of Montpellier
Montpellier
France

Nathalie Debese
Ocean Sensing and Mapping
ENSTA Bretagne
Brest Cedex 09
France

Isabelle Quidu
Ocean Sensing and Mapping
ENSTA Bretagne
Brest Cedex 09
France

Benoît Clement
Ocean Sensing and Mapping
ENSTA Bretagne
Brest Cedex 09
France

Annick Billon-Coat
Pôle STIC
ENSTA Bretagne
Brest Cedex 09
France

ISSN 2194-6396   ISSN 2194-640X   (electronic)
Ocean Engineering & Oceanography
ISBN 978-3-319-32105-9   ISBN 978-3-319-32107-3   (eBook)
DOI 10.1007/978-3-319-32107-3

Library of Congress Control Number: 2016938653

Printed on acid-free paper

This Springer imprint is published by Springer Nature
The registered company is Springer International Publishing AG Switzerland

Kees de Jong (1962–2014)
This book is dedicated to the memory of Kees de Jong, Research and Development manager for Fugro Intersites BV.
The organizing committee of the conference MOQESM'14 is deeply saddened to learn the death of Kees de Jong, Visiting Professor in the session hydrography of MOQESM, tragically died afterwards the conference.
Kees de Jong contributed to the development of Fugro Research activities since 2003, first as senior geodesian and then, from 2007, as Geodesy Department Manager in 2007. He formerly worked as Assistant Professor with Delft University in the high accuracy positioning domain. As an attempt to put industry closer to academics, he fostered much collaboration between some companies and some academic institutes as Newcastle University (UK) or ENSTA Bretagne (France) for instance.
All collaborators, whether colleagues or students gave him a deep respect as much due to for his scientific skills, his commitment to applied research and industry, as for his human quality. For all of us, Kees will remain a model both as human and as scientist.

Preface

Every 2 years, MOQESM is organized in Brest during the Sea Tech Week with the aim to focus on emergent techniques for quantitative monitoring of the underwater environment; MOQESM standing for MOnitoring Quantitatif de l'Environnement Sous-Marin. The 2014 edition of the conference, MOQESM'14, is the opportunity for people of the research and industry communities to meet, attend, and discuss with specialists of two research domains: marine robotics and coastal hydrography, with application to the coastal environment mapping and the survey of underwater infrastructures. The objective of the MOQESM'14 conference is to demonstrate that, though being very distinct, the two domains of marine robotics and coastal hydrography can take benefit from research progress in each other, in the future, in order to design new products and mapping methods combining them. The recent research and industrial achievements in these two domains are developed in the 11 papers gathered into the proceedings of MOQESM'14. The conference is organized in two plenary sessions headed with invited talks.

The first chapter of this book is dedicated to the improvements in hydrography. It begins with an invited talk given by Carole Nahum, from the Délégation Générale pour L'Armement (DGA), about defense needs and strategies in terms of environment monitoring. Techniques to acquire the underwater environment can be improved in many ways: from the positioning accuracy to the fusion of multiple sensors. Five scientific contributions constitute this first chapter. Precise mapping of the underwater environment requires accurate positioning of the acquired data. New approaches to obtain an accuracy of a few centimeters rely on Global Navigation Satellite Systems (GNSS). To reach such accurate positioning, Kees de Jong et al. propose an approach based on merging PPP techniques (use of precise satellite orbits and clocks) with Integer Ambiguity Resolution (IAR), known from GNSS Real-Time Kinematic (RTK) positioning techniques. As the accuracy of the acquired bathymetric data also depends on the motion of the sensors, Nicolas Seube, Sebastien Levilly, and Kees de Jong present an automatic method to estimate the angular alignment between the Inertial Measurement Unit (IMU) and the multibeam echo sounder. Acquiring the bathymetry can become a very difficult task when the environment is challenging and not cooperative: high-flowing rivers, confined zones and ultra-shallow waters. In such environment, unreachable with conventional survey launches, Mathieu Rondeau et al. proposed an autonomous drifting buoy equipped with a GNSS receiver, an IMU, and a single-beam echo sounder. The acquisition and the monitoring of the underwater environment can be improved by combining different sensors. Claire Noel et al. present new tools to produce operational seabed maps by fusing the information collected by several acoustic systems operating simultaneously or not. This session concludes with the higher level issue on how to efficiently make available the data from the marine environment to end-users like marine industries, decision-making bodies, or scientific research. As in Europe the marine data are stored in a wide range of national, regional, and international databases and repositories using different formats and standards, J.-B. Calewaert et al. present the European Marine Observation and Data Network (EMODnet). EMODnet is a network of organizations set up in 2007 by the European Commission in the framework of EU's Integrated Maritime policy to address the fragmented marine data collection, storage and access in Europe.

The second chapter addresses new developments in marine robotics. The first invited speaker, Edson Prestes from Universidade Federal do Rio Grande do Sul, Porto Alegre, Brasil, proposes a new approach to the global positioning of underwater robots based on probability and interval analysis. The second invited speaker, Vincent Rigaud, IFREMER, France, introduces a new kind of underwater robots resulting from the hybridization of a Remotely Operated Vehicle (ROV) with an Autonomous Underwater Vehicle (AUV). Marine robotics has to operate in the very challenging oceanic environment. To design and build effective robots, a wide range of research topics must be addressed, e.g., underwater communication, obstacle avoidance, software design for embedded systems, control/command, sensor design and integration, algorithms for autonomous navigation, localization, and positioning. Below the sea surface high-frequency electromagnetic communication shows poor performance and acoustic waves are preferred. However, acoustic modems generally remains costly for small robots and Christian Renner et al. have studied a new acoustic modem design aimed at low power consumption, small form factor, and low unit cost. Before addressing the robot itself, it is important to improve the sensing devices. As for communications, the preferred technique for imaging the seabed is based on ultrasonic acoustic waves. New techniques based on synthetic aperture, multiple aspects and interferometry allow for both accurate measurement of the bathymetry and optics-like imaging of the sea floor. Myriam Chabah et al. present the design and discuss the first experimental results of the SAMDIS sonar system which first implements simultaneously these new techniques. Another scientific challenge is to efficiently design the code executing autonomous mission. Such code has several levels of abstraction from low-level control loops to high-level path planning. Goulven Guillou and Jean-Philippe Babau have developed IMOCA; a generic multi-platform model-based approach to code generation for embedded systems. At low level, the efficient control of a robot can be achieved by taking into account the hydrodynamics of the robot. Yang Rui et al. present this approach and apply it to the

Ciscrea AUV. At intermediate level, the AUV can be controlled using its vision sensor. Eduardo Tosa et al. show how visual servoing implemented in Coralbot AUV solve the problem of detecting coral reef. At higher level, when navigating on the surface or underwater, autonomous robots have to find a safe path. For example, autonomous navigation of a surface vessel must take care of the shore line and the other vessels. To solve this problem Silke Schmitt et al. proposed a vector field approach.

The main conclusion of MOQESM'14 is that, although different and often separate, the domains of marine robotics and hydrographic measurements share some research topics like global positioning, acoustic sensing, data processing, or mission planning. The content of this book also demonstrates that marine robotics will play an increasing role in acquiring the marine environment.

Benoît Zerr

Head of "Systems of Drones" Program

Lab-STICC

Organizing Committee

ENSTA Bretagne, Lab-STICC UMR CNRS6285

Luc Jaulin, Isabelle Quidu, Benoît Zerr, Benoît Clement, Nathalie Debese, Annick Billon-Coat

GDR Robotique

Vincent Creuze

CIDCO—Interdisciplinary Center for the Development of Ocean Mapping, Canada

Coralie Monpert, Nicolas Seube, Jean Laflamme

Invited Speakers

Pr Edson Prestes
Leader of Phi Robotics Research Group at UFRGS
Federal University of Rio Grande do Sul, Porto Alegre, Brésil

Carole Nahum
Manager Domain Environment and Géosciences of Directorate for the Armed Forces of the French MoD

Vincent Rigaud
Head of the Underwater Systems Unit, IFREMER Toulon

Chairmen

Benoît Zerr, ENSTA Bretagne
Hervé Bisquay, GENAVIR
Vincent Creuze, GDR Robotique

Contents

Introduction

The French Directorate for the Armed Forces, commonly named DGA *Direction Générale pour l'Armement* belongs to the Ministry of Defense. It is in charge of equipping the navy, the army, or the air forces with devices such as sensors or weapons, new vehicles such as ships, aircraft carriers, underwater autonomous vehicles, or submarines, and of maintaining them in working order. When conducting operation, a precise positioning system, a navigation unit, data transmission or communication devices, sensors for detection, and tracking or recognition of targets are needed.

Furthermore, DGA is in charge of proposing software for decision-making or mission planning. Knowing the environment, which means the physical state of the atmosphere, the land, or the sea, and being able to forecast the changes and the dangerous events that could happen is a challenge and a prerequisite for these tools.

Thinking about the future (within 10–20 years), one must take into account not only political and strategic changes over the world but also progress of scientific research or technologic improvements. It is our responsibility to incorporate them into the devices. Therefore, DGA supports scientific studies and technological projects proposed by laboratories or SMEs respectively, connected to the needs of the forces. But on the other hand, researchers and designers must cope with some constraints such as integration of sensors on small platforms (constraints of weight, size, energy supply, etc). Robust hardware, especially in hostile areas, must be designed since electronic devices may be damaged by particular environmental conditions or their performances drastically reduced. The algorithms must also fulfill several requirements such as real-time running. This is particularly challenging when conducting or planning activities in the ocean. The underwater environment is not accessible via satellite or airborne sensors (optical, infra-red, or RADAR) and specific technics must be addressed.

For these reasons, the International Conference on Quantitative Monitoring of Underwater Environment (MOQESM) which gathers researchers and SME's designers offers a wide range of topics of great interest for DGA. In particular, the session "Hydrography: from sensors to products" presents smart devices, new methods, and algorithms.

The military in operation needs up-to-date and precise information. For example, positioning in the underwater is crucial. As for land operations, a map of the bottom of the ocean is the basic tool. How to get it? Bathymetry is usually deduced from measurements of gravity and its derivatives. So even if we develop small gravimeters, this would require the visit of an area and would take time to draw the map. Unfortunately, we often have to manage in some unknown regions. Among military activities, one must not forget survey for which several types of sensors may be used in order to retrieve bathymetry but also the nature of the seabed. It is very relevant for us to propose some real-time technics.

Interferometric sonars are powerful tools for shallow water survey. Unfortunately they suffer uncertainties which may degrade bathymetry quality. The paper "Real-time sounding uncertainty estimation in phase measuring bathymetric sonars" presented by Kongsberg GeoAcoustics Ltd. develops a method for calculating in real time the uncertainties of a commercial PMBS.

In every activity, the military, as anybody, has to keep safe and to care about their impacts on ecosystems. Mapping is also used for finding mines which can be dangerous or putting small devices for survey. In order to draw a map of the seabed, SEMANTIC (France) proposes to integrate several types of acoustic sensors on a small ship and to develop a new data fusion method in "New tools for seabed monitoring using multi-sensors data fusion." This is particularly relevant for us since this can be performed with very low-cost sensors and nearly in real-time.

Improving offshore positioning in real time using Global Navigation Satellite System (GNSS), in particular for tidal applications, may be challenging. The usual accuracy is 3–5 cm horizontally and 6–10 cm vertically. The problem is addressed in the paper "New developments in precise offshore GNSS positioning" proposed by Fugro Intersite. Merging PPP technics (precise satellite orbits and clocks) with Integer Ambiguity Resolution (IAR) allows a better accuracy.

Accuracy relies on a boresight calibration between IMU and Multibeam Echo Sounder (MBES). ENSTA Bretagne (France) and CIDCO (Canada) propose in their paper "Automatic boresight calibration of hydrographic survey systems," a multi-dimensional optimization concept which should provide statistical analysis to be integrated in every calibration report

Deployment of AUVs equipped with several types of sensors such as DVL, pressure sensor, sound velocity sensor, long base line systems, and Inertial Navigation System (INS) in order to collect information is a challenge since the autonomous retrieval of the position may be erroneous. XBLUE (France) proposes in "Optimizing survey deployment and processing times using sparse LBL positioning," a new concept of sparse array navigation and prove that an optimized coupling between inertial unit and the acoustic positioning system may result in a decimetric positioning precision.

CIDCO (Canada) introduces the Hydroball system which is an autonomous drifting buoy equipped with a GNSS receiver, an IMU and a single-beam echo sounder, for surveying hostile and non-accessible areas, in particular ultra-shallow waters. This system is shown to meet industrial international hydrographic standards.

Finally in order to elaborate a "picture" of the marine environment, clear enough, pertinent and faithful, it is necessary to collect data, qualify and interpret it. This would help for understanding physical phenomena, modeling and characterizing the different areas and their spatiotemporal evolution by assimilation of data. The European Marine data and Observation Network (EMODnet) is a gateway for marine and coastal data. It is a network of organizations (set up in 2007 by the European Commission) in charge of addressing the fragmented marine data collection, storage and access in Europe.

As a conclusion, let us mention that several other topics beyond the scope of MOQESM are also of great interest for DGA. Among them, knowledge of coastal environment (beaches change), improvement of bathymetry resolution, sedimentology relevant when looking for a place on the seabed where to locate a device, geoacoustics in order to tune sonars, turbidity for search and rescue, courants and tides, underwater sound transmission, ice monitoring and so forth.

Carole Nahum
Manager Domain Environment and Geosciences—DGA/MRIS

Part I
Hydrography: From Sensors to Products

New Developments in Precise Offshore GNSS Positioning

Kees de Jong, Matthew Goode, Xianglin Liu and Mark Stone

Abstract Global Navigation Satellite System (GNSS) based Precise Point Positioning (PPP) has become the de facto standard for precise real-time offshore positioning applications. Current precision is of the order of 3−5 cm horizontally and twice this value for the vertical. However, this may not yet be good enough for tidal applications. In this contribution we will discuss new developments at Fugro, one of the world's main providers of precise offshore real-time GNSS positioning services, to further improve PPP precision to the 2−3 cm level in the vertical component worldwide. For demanding applications, it is possible to even further improve this precision. These developments are based on merging PPP techniques (use of precise satellite orbits and clocks) with Integer Ambiguity Resolution (IAR), known from GNSS Real-Time Kinematic (RTK) positioning techniques. PPP IAR requires the generation and distribution to mobile users of Uncalibrated Phase Delays (UPDs) a network of reference stations. The network can be as small as one station or cover the entire globe. Once applied to the data of a mobile receiver, the carrier ambiguities should be integer. Fixing of these ambiguities to their proper integer value will result in significantly improved positioning performance. The infrastructure used to generate precise orbits, clocks and UPDs will be discussed. PPP IAR results will be shown from regional and global test beds, based on Fugro's precise orbits and clocks for all currently available GNSSs. In addition, it will be shown that the introduction of new systems and signals, like triple-frequency GPS, Galileo and BeiDou, will help to significantly reduce the time required for PPP IAR solutions to converge to this centimeter level of accuracy.

Kees de Jong—Deceased

K. de Jong · M. Goode (✉) · X. Liu (✉) · M. Stone
Fugro Intersite B.V., Dillenburgsingel 69, 2263HW Leidschendam, The Netherlands
e-mail: M.Goode@fugro.nl

X. Liu
e-mail: x.liu@fugro.nl

B. Zerr et al. (eds.), *Quantitative Monitoring of the Underwater Environment*,
Ocean Engineering & Oceanography 6, DOI 10.1007/978-3-319-32107-3_1

1 Introduction

Fugro provides a number of GNSS services for precise offshore positioning. The current services are summarized in Table 1. Main distinction is between differential (L1, HP) and PPP (XP, G2).

The infrastructure consists of about 150 reference stations. One reference network of about 100 stations is used for the differential services, another, consisting of 45 stations, for PPP services. The PP network is used to generate precise GNSS orbits and clocks in real-time. Currently this is done for GPS, Glonass and BeiDou. Once Galileo is declared operational, this system will immediately be included as well. It should be noted that Fugro was the first provider to show real-time PPP results from a Galileo-only solution, already in March 2013, see [5].

The main reason for providing both differential and PPP services is independence. If one type of positioning fails, it should be possible to continue with another. This is also the reason why the two networks are separated. There is also redundancy in the Network Control Centers (NCCs), with one in Houston and the other in Perth and the number of data links used to broadcast GNSS correction data to users in the field. Figure 1 shows Fugro's GNSS augmentation infrastructure.

The G2 service currently provides a positioning accuracy of about 3−5 cm horizontally and twice this value for the vertical (one sigma). However, for tidal applications this precision is not good enough. Therefore, work is going on to further improve precision by blending PPP with Integer Ambiguity Resolution (IAR) known from RTK (Real-Time Kinematic) techniques, which are mainly used on land and using dense reference networks, or, in other words, using short baselines between reference and mobile stations (usually up to 100 km). For offshore applications, the baselines are much longer, roughly up to 1000−1500 km, and IAR becomes more complex, as e.g. atmospheric effects can no longer be ignored or constrained and have to be explicitly taken into account. As a result, convergence times (the time is takes to reach (sub-)dm accuracy) are much longer than for standard RTK techniques.

Table 1 Fugro GNSS positioning services

Service	Accuracy	Correction source	Navigation satellites	Signal frequencies	Positioning mode
G2	Decimeter	Orbit and clock	GPS, Glonass, BeiDou	Dual	PPP
XP	Decimeter	Orbit and clock	GPS	Dual	PPP
HP	Decimeter	Reference stations	GPS	Dual	Differential
L1	Meter	Reference stations	GPS	Single	Differential

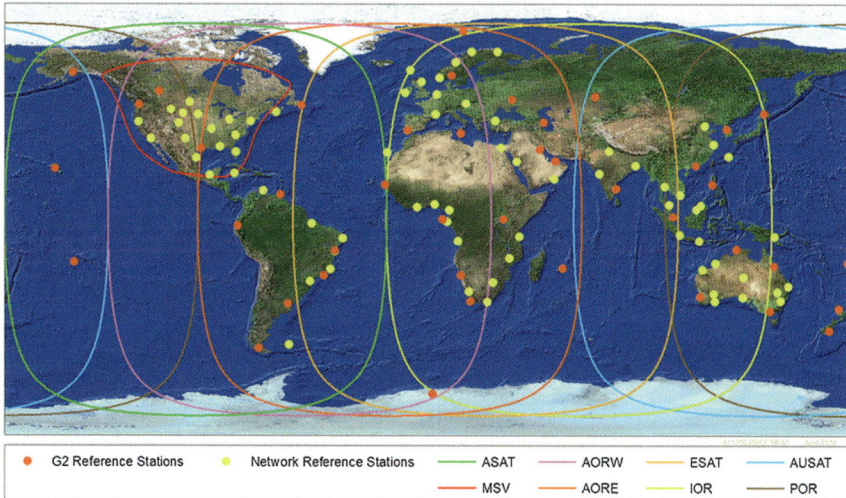

Fig. 1 GNSS augmentation structure: reference stations and footprints of satellite beams

2 PPP and PPP IAR

PPP uses GNSS code (pseudo range) and carrier observations to estimate a mobile user's precise position, see e.g. [6, 11]. Other parameters that are estimated are receiver clock biases, atmospheric effects and carrier ambiguities.

When positioning a mobile using PPP, it does not explicitly use reference stations, as RTK does (implicitly it does, as reference stations are used to generate precise satellite orbits and clocks). A consequence of not using reference station data is that the estimated carrier ambiguities refer to a single station (the mobile) and therefore are not integer. Only double difference ambiguities (ambiguities referring to two stations and two satellites) are integer. If the integer ambiguities can be estimated values, they can be held fixed in a subsequent adjustment, resulting in a much more precise position estimate. In order to make ambiguities integer, additional corrections are needed, estimated from reference stations. These reference stations can be the same as the ones used for generating orbit and clock parameters, but this is not required. As long as the reference network uses the same orbits and clocks as the mobile, it is fine. A reference network can be as small as a single station, but more are preferred to increase redundancy, improve precision and extend the coverage area.

The estimated corrections, usually referred to as Uncalibrated Phase Delays (UPDs) or Fractional Carrier Biases (FCBs), should be applied to the carrier observations of a mobile station data in order to make the ambiguities integer. This shows that PPP with Integer Ambiguity Resolution (PPP IAR) is in fact a differential technique. It should also be emphasized that the integer ambiguities which are estimated, are double difference ambiguities, but parameterized in an undifferenced mode, see e.g. [2]. In recent years, a number of PPP IAR methods were developed, such as the ones described in [1, 4, 7, 10].

3   PPP and PPP IAR Tests

In order to test the concept of PPP IAR, we performed a real-time and an offline test.

3.1   *Real-Time Test*

For the real-time test, we used G2 orbits and clocks and Fugro reference stations from the HP network in North America to generate UPDs. These UPDs were then applied to reference stations that were not used to generate UPDs. Three weeks of data, with an observation interval of one second was processed in kinematic mode, i.e., independent positions were computed for each observation epoch. This is a realistic scenario as data may be missing in the real-time data stream, which may affect the generation of UPDs.

The reference station network is shown in Fig. 2. Shown in Figs. 3 and 4 are PPP and PPP IAR results for station H332 for the entire period. Results for the other station, H372, were very similar. It can be clearly seen that IAR has a significant impact on the precision of the positioning results, with standard deviations more or less improved by a factor two.

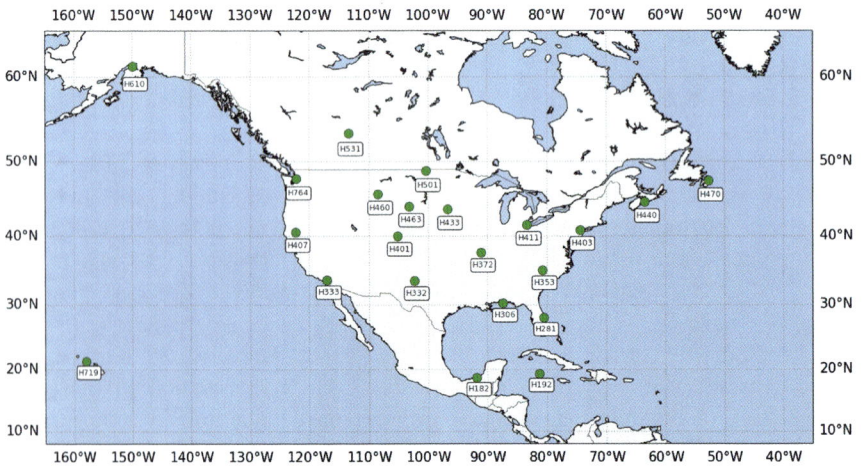

Fig. 2 Reference stations used to generate one set of UPDs for the real-time test. Stations H332 and H372 were considered as mobiles and not used for the UPD generation

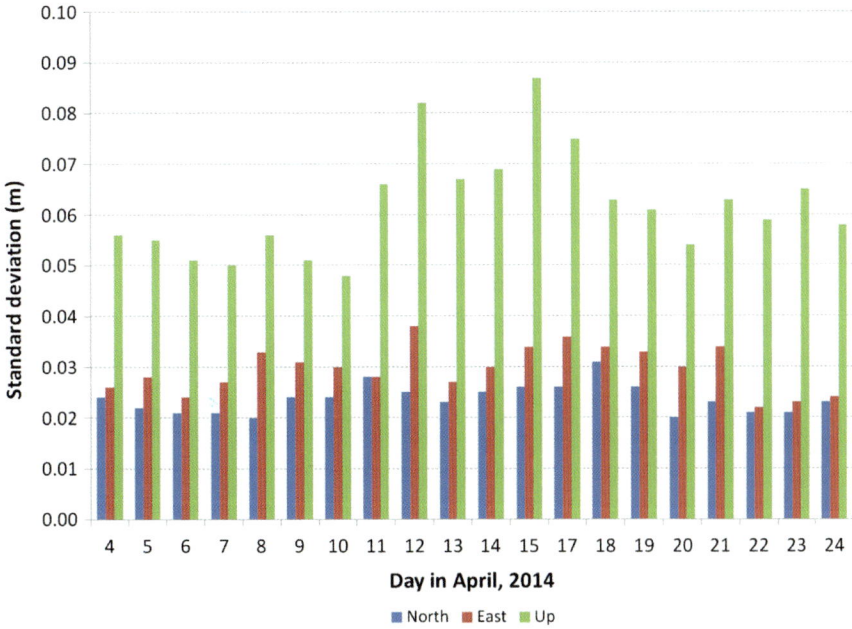

Fig. 3 PPP results for station H332 for the three week real-time test period

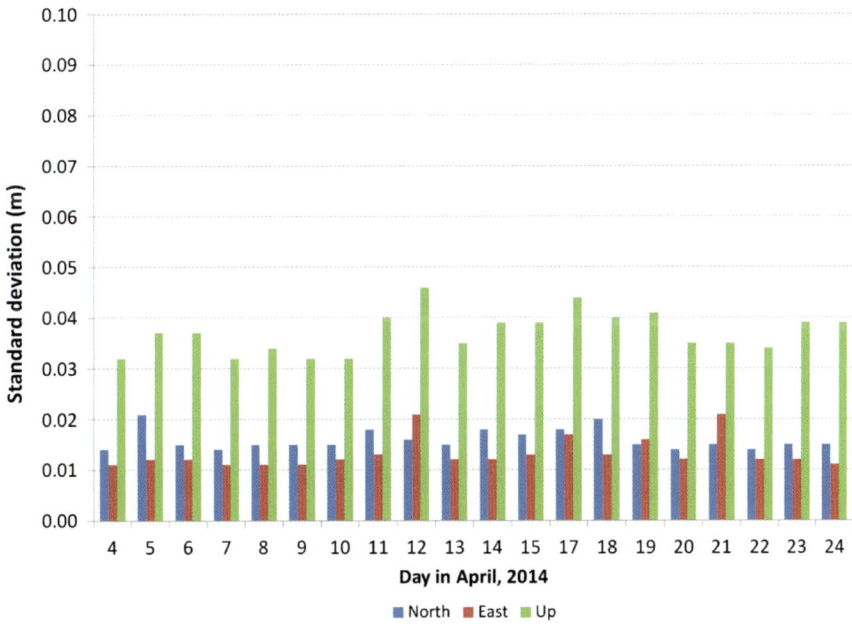

Fig. 4 PPP IAR results for station H332 for the three week real-time test period

3.2 Offline Test

For the offline test, we again used G2 orbits and clocks, but this time we used observation data from the US CORS (Continuously Operating Reference Station) network to generate UPDs and for mobile position estimation. Again, stations that were used for the positioning part were not used to generate UPDs. Figure 5 shows the 20 reference stations used to generate UPDs, Fig. 6 the 111 stations used for

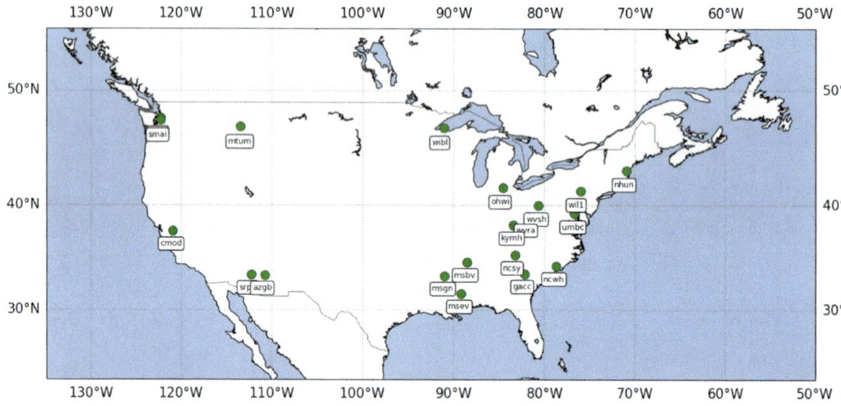

Fig. 5 US CORS stations used to generate one set of UPDs for the offline test

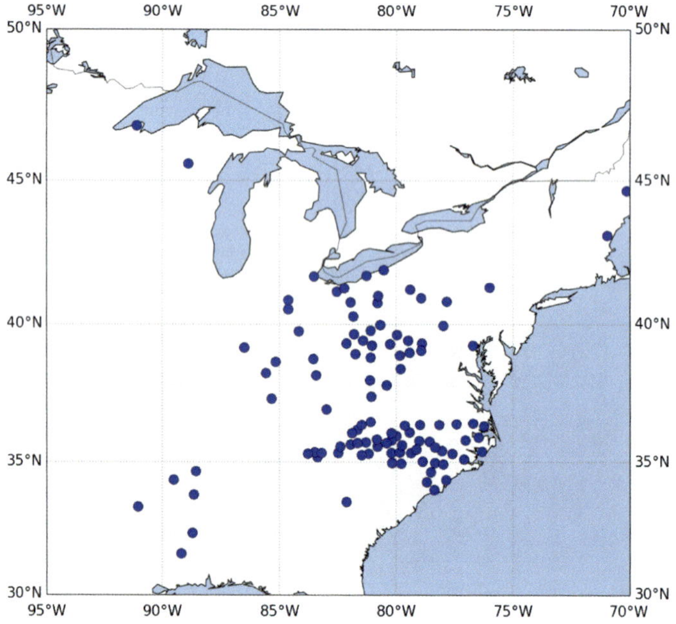

Fig. 6 Location of 111 stations used as mobiles in the offline test

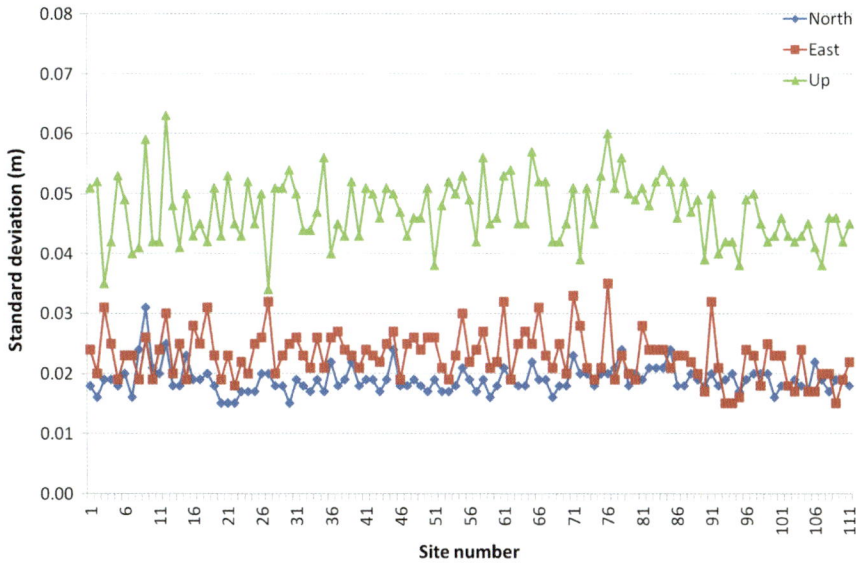

Fig. 7 PPP results for the 111 mobile stations of Fig. 6 for one day

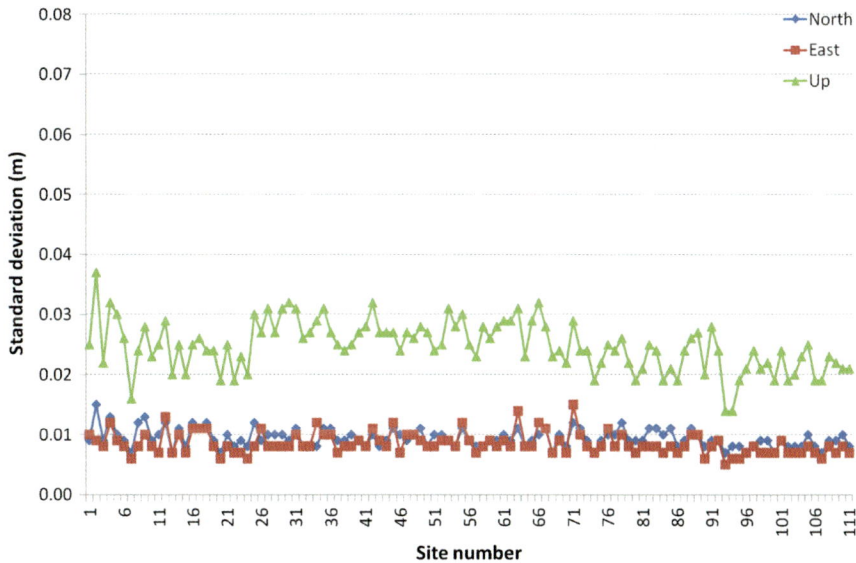

Fig. 8 PPP IAR results for the 111 mobile stations of Fig. 6 for one day

kinematic positioning. There were no gaps in the data. Observation interval was one second. As can be seen from Figs. 7 and 8, the PPP and PPP IAR results are slightly better than for the real-time case. This is mainly due to improved data delivery. For both tests the same real-time orbits and clocks were used.

4 Convergence Time

The ambiguity success rate, [9] is defined as the probability of fixing the carrier ambiguities to their correct integer values. It requires the covariance matrix of the float ambiguities, but no actual data. It is therefore a design parameter, just like the popular DOP (Dilution Of Precision) values.

In the example shown here, we defined a minimum success rate of 99.9 % and then compute the number of epochs required to reach this success rate. It is assumed satellite geometry does not change, so once we have computed the geometry, we only increase the number of epochs in order to improve precision of the estimated float ambiguities. Next, the ambiguities are decorrelated using the LAMBDA method, [3, 8] after which the success rate can be computed from the variances of the decorrelated ambiguities.

Once the ambiguities are fixed to their integer values, the carrier phase measurements become very precise pseudo ranges. With these precise observables, it is straightforward to compute a precise position.

As we know from standard RTK applications, even though the float solution is not very accurate (in other words, it is not converged to something precise), it is often already possible to estimate the integer values of the ambiguities, resulting in a very precise ambiguity fixed solution after a small number of epochs.

The same is possible with PPP IAR: ambiguities can be fixed while the solution has not yet converged. As a result, convergence time can be reduced, often significantly, as shown by the GPS only and GPS/BeiDou example in Figs. 9 and 10 for a station in China (where BeiDou satellite visibility is much better than in Europe).

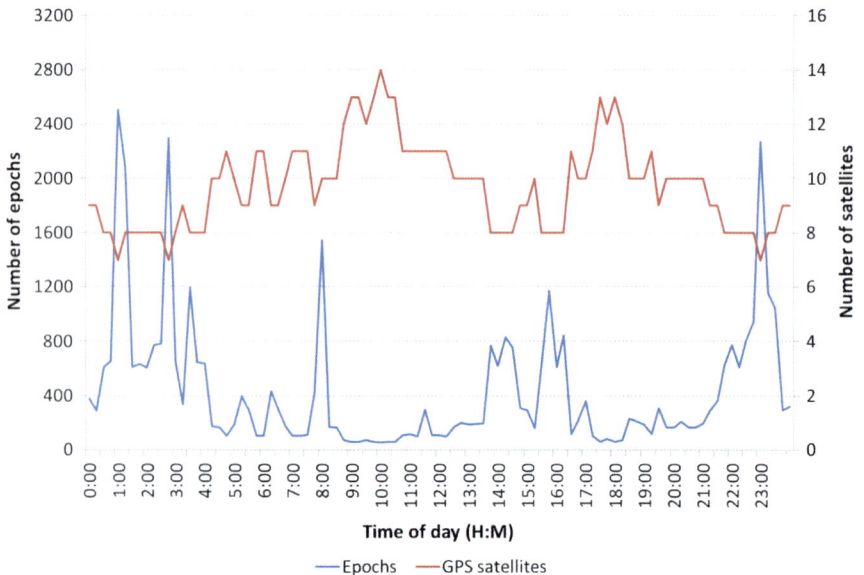

Fig. 9 Number of epochs required for an ambiguity success rate of 99.9 % for a station in China using dual-frequency GPS only

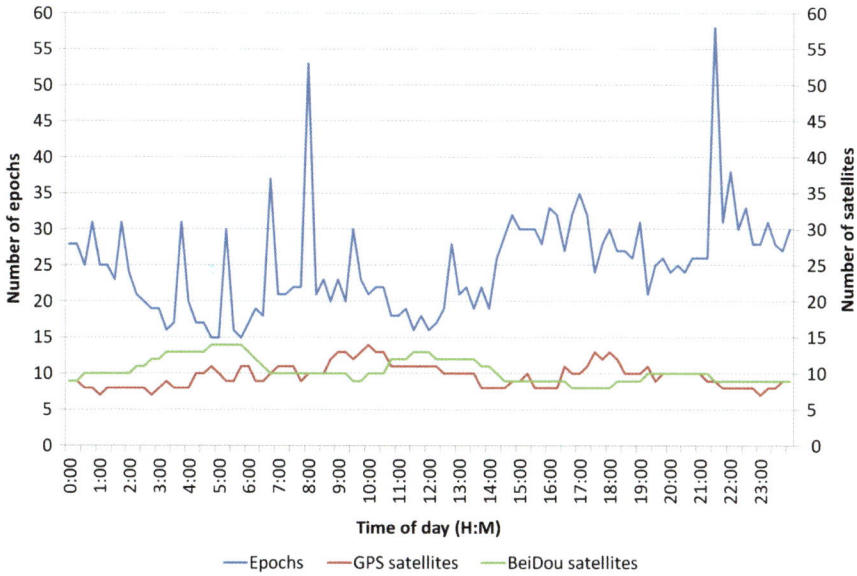

Fig. 10 Number of epochs required for an ambiguity success rate of 99.9 % for a station in China using dual-frequency GPS and triple-frequency BeiDou (based on the current (September 2014) constellation of 14 satellites)

As we can see from this figure, the number of epochs required to fix ambiguities for a GPS only solution is in general quite high. Adding the current constellation of BeiDou satellites significantly reduces the number of epochs required to fix ambiguities and therefore convergence time. Note that this reduction in convergence time does not require any external information, such as ionospheric data. The GNSS observations themselves provide enough information to converge to cm level precision in less than a minute.

5   Conclusions

PPP IAR is feasible using Fugro's real-time G2 orbits and clocks and a sparse network to compute UPDs. Compared to standard PPP, precision improves by a factor two. With vertical accuracies of the order of 2–3 cm, it becomes feasible to use GNSS for tidal applications. Using smaller networks to compute UPDs, together with atmospheric information from these networks, may further improve precision down to the standard RTK level.

Using multiple GNSSs, such as GPS and BeiDou, will significantly reduce the time required to fix ambiguities and therefore convergence time to reach cm level accuracy in PPP IAR.

References

1. Collins, P., Lahaye, F., Héroux, P., & Bisnath, S. (2008). Precise point positioning with ambiguity resolution using the decoupled clock model. *Proceedings ION GNSS 2008*, 1315–1322.
2. de Jonge, P.J. (1998). *A processing strategy for the application of the GPS in networks*. Ph.D. thesis, Delft University of Technology, pp. xi–225.
3. de Jonge, P.J., Tiberius, C.C.J.M. (1996). The LAMBDA method for integer ambiguity estimation: implementation aspects. *Publications of the Delft Geodetic Computing Centre, LGR series*, 12, pp. v–49.
4. Ge, M., Gendt, G., Rothacher, M., Shi, C., & Liu, J. (2008). Resolution of GPS carrier-phase ambiguities in precise point positioning (PPP) with daily observations. *Journal of Geodesy, 82*, 389–399.
5. GPS World (2013). *Real-time PPP with Galkleo demonstrated by Fugro*. GPS World, Mar 2013, http://gpsworld.com/real-time-ppp-with-galileo-demonstrated-by-fugro/. Accessed 16 Sept 2014.
6. Kouba, J. (2009). A guide to using International GNSS Service (IGS) products. International GNSS Service, pp. 34.
7. Laurichesse, D., & Mercier, F. (2007). Integer ambiguity resolution on undifferenced GPS phase measurements and its application to PPP. *Proceedings ION GNSS, 2007*, 839–848.
8. Teunissen, P. J. G. (1995). The least-squares ambiguity decorrelation adjustment: A method for fast GPS integer ambiguity estimation. *Journal of Geodesy, 70*, 65–82.
9. Teunissen, P. J. G. (1998). Success probability of integer GPS ambiguity rounding and bootstrapping. *Journal of Geodesy, 72*, 606–612.
10. Teunissen, P. J. G., Odijk, D., & Zhang, B. (2010). PPP-RTK: Results of CORS network-based PPP with integer ambiguity resolution. *Journal of Aeronautics, Astronautics and Aviation, Series A, 42*, 223–230.
11. Zumberge, J. F., Heflin, M. B., Jefferson, D. C., Hefkins, M. M., & Webb, F. H. (1997). Precise point positioning for the efficient and robust analysis of GPS data from large networks. *Journal of Geophysical Research, 102*(B103), 5005–5017.

Automatic Estimation of Boresight Angles Between IMU and Multi-Beam Echo Sounder Sytems

Nicolas Seube, Sébastien Levilly and Kees de Jong

Abstract Nowadays, the boresight calibration between IMU (Inertial Measurement Unit) and MBES (Multi-Beam Echo Sounder) systems is achieved by the patch-test procedure which estimates the three boresight angles: roll, pitch and yaw. That procedure consists in two steps. The first one is the selection of an overlapping area. That selection is done thanks to the experience of a surveyor. The second step evaluates the roll, pitch and yaw angles separately by a method which tries a subset of possible angles. For theses possible angles, the discrepancy between digital terrain models (one DTM by survey line) is calculated in the previous selected area and the minimum is assumed to be the "optimal" solution. This paper presents some preleminary results from a research project between FUGRO, ENSTA Bretagne and CIDCO. This project aim is to design new methods in the calibration topic. These procedures use multi-dimensional optimization concepts in order to provide statistical analysis which should appear in any calibration report.

1 Introduction

We consider the problem of boresight calibration of a hydrographic system, composed by a Multi-Beam Echo Sounder (MBES), an Inertial Measurement Unit (IMU) and a positioning system (generally being a GNSS receiver). This hydrographic system, as all mobile mapping systems, enables one to determine the position of soundings in a geographic frame from the knowledge of raw source data from the MBES,

---

Kees de Jong—Deceased

N. Seube (✉)
CIDCO, Rimouski, Canada
e-mail: nicolas.seube@cidco.ca

S. Levilly
ENSTA Bretagne, Brest, France
e-mail: sebastien.levilly@ensta-bretagne.org

K. de Jong
Fugro Intersite B.V., Leidschendam, The Netherlands

© Springer International Publishing Switzerland 2016
B. Zerr et al. (eds.), *Quantitative Monitoring of the Underwater Environment*,
Ocean Engineering & Oceanography 6, DOI 10.1007/978-3-319-32107-3_2

IMU and GNSS receiver. This can be done by using a spatial referencing equation and a simplified version is the following Eq. (1).

$$\mathbf{X}_n(t) = \mathbf{P}_n(t) + C_{bI}^n(t - dt)[C_{bS}^{bI}\mathbf{r}_{bS}(t) + \mathbf{a}_{bI}] \tag{1}$$

where $\mathbf{X}_n = (x, y, z)_n$ is the position of a sounding in a navigation frame *(n)* (which can be a local geodetic frame), $\mathbf{P}_n$ is the position delivered by the GNSS receiver in frame *(n)*, C_{bI}^n is the coordinate transformation from the IMU body frame to the navigation frame (which can be parametrized using Euler angles (φ, θ, ψ), denoting roll, pitch and yaw, respectively), the MBES return r_{bS}, coordinated in the MBES frame *(bS)*, the lever-arm vector coordinated in the IMU frame a_{bI} and the boresight coordinate transformation C_{bS}^{bI}.

In Eq. (1), t denotes the reference time from the GNSS, which is supposed to be propagated to the IMU through a distributed time and message synchronization system [1], and dt denotes a possible latency between the MBES and the IMU.

The dependency of the calibration parameters on soundings spatial referencing is described by Eq. (1), among them are:

- dt, the latency between the IMU and the MBES system (it is to be noticed that in most modern hydrographic systems, latency between GNSS and the MBES impact can be considered as negligible, but latency between the MBES and IMU is not [9];
- C_{bS}^{bI}, the boresight coordinate transformation;
- $\mathbf{a}_{bI}$, the lever-arms which may be affected by static measurement errors, coordinate transformation errors from the measurement frame to the IMU frame, and in some cases, time-varying (for large ships for instance);
- The MBES range and beam launch angles, affecting the term $\mathbf{r}_{bS}$.

This article will focus on the estimation of boresight coordinate transformation C_{bS}^{bI}, as an essential component of calibration parameters. A classical method to determine this transformation is the so-called "patch-test" which principle is briefly recalled here and which limitations are hereafter detailed. The patch-test decouples the three boresight angles estimation problem, and starts with the roll angle, followed by the pitch, and then the yaw angle. For the roll angle, a flat bottom, surveyed in opposite direction is used, since the roll boresight $\delta\varphi$ effect can be easily characterized (see Fig. 1).

We illustrate the pitch boresight calibration method, which uses nadir data from two opposite lines over a slope. Figure 2 illustrates the effect of a pitch boresight $\delta\theta$ over a regular slope, followed by a flat terrain.

The estimation of the yaw boresight is classically done by identifying a target over a flat bottom, and by surveying this target using two lines in the same direction, with outer beams intersecting the target. In order to improve the resolution of this method (which may suffer from the fact that the target size may not be significant enough, which causes uncertainty on the tie point position precision and accuracy), one can also survey two parallel lines in the same direction over a regular slope.

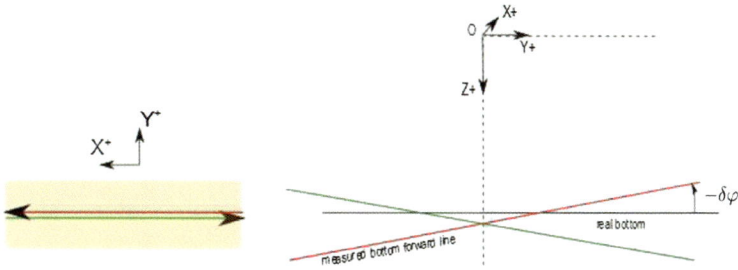

Fig. 1 Effect of roll boresight and two opposite lines on a flat seafloor. Two lines are surveyed in opposite directions; The angle between the two lines is $2\delta\varphi$. In a typical roll boresight estimation problem, the 95 % confidence interval of the estimated roll boresight that can be achieved by using a characterization method (such as estimating the angle between fitted lines or planes in the overlapping area) is about 0.05°, which is much worse than the precision of a tactical grade IMU

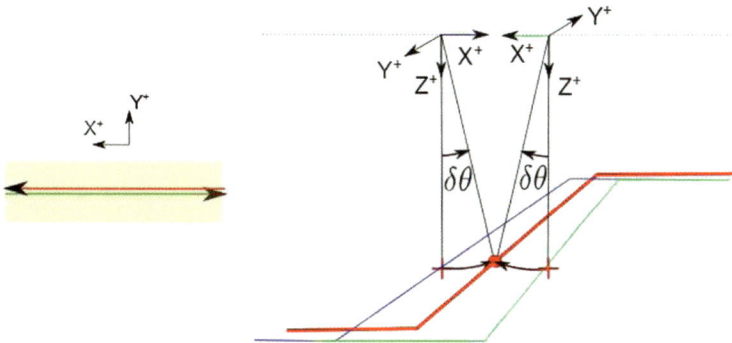

Fig. 2 Effect of pitch boresight on two opposite lines over a slope

In most data acquisition software, boresight estimation is achieved by two steps:

1. The user selects a subset of the data set from overlapping areas;
2. In reviewing all the possible boresight values over a given interval, it re-computes corrected data from source data according to the spatial referencing Eq. (1), and builds a digital terrain model (DTM) for the different overlapping data sets. Then, it compares the discrepancy between those models, and chooses the lowest one.

We observe that through this process, the choice of the analysis area dramatically impacts the boresight estimation, and is left to the user. Secondly, these methods are not properly based on optimization methods, since they massively compute all possible values of the corrected DTM for all possible values of the roll angle, the pitch angle and then the yaw angle. Consequently, they cannot deal with the problem of coupling between angles, since a true 3D computation in the boresight angles space is not achievable. Moreover, the patch-test procedure does not provide any estimate of the boresight precision, which would be highly desirable.

The effect of boresight on survey data is rather complex, with each swath being modified according to the local seafloor morphology which determines the beams grazing angles and therefore impacts the error between the actual and assumed sounding. Indeed, a boresight error acts as a rotation around the acoustic center of the MBES.

From a global point of view, the effect of boresight acts as a rotation around a time varying center (i.e., the position of the MBES). It is therefore impossible to model the effect of the boresight angle over a global surface by a simple geometric transformation like a similarity transformation for instance. In Fig. 2, the nadir beams are plotted for two opposite survey lines over a slope and flat areas. From this figure, it can be easily seen that it is impossible to deduce the actual sea floor from the assumed seafloors by a simple geometric transformation. From this remark, we deduce that boresight must be determined from a local analysis.

Another problem is the coupling between roll, pitch and yaw angles, which can be understood from the spatial referencing Eq. 1. Indeed, entries of the coordinate transformation matrix (in NED convention) $C_{bS}^{bl} = C_3(\delta\psi)C_2(\delta\theta)C_1(\delta\varphi)$ depend on the three boresight angles, which means that they contribute to each swath return distortion by coupling. We have seen that the classical patch-test method first determines the roll, then the pitch and finally, the yaw boresight. This implies that the roll boresight is determined with uncorrected pitch and yaw. In case of a non-perfectly flat sea-floor, pitch and yaw actually contribute to the MBES swath return distortion. This effect of boresight angles cross-talk has the following consequence. The determination of roll is biased by the absence of knowledge of pitch and yaw which impact data used for roll calibration over non-perfectly flat local surfaces. After roll determination, the pitch is estimated using nadir data over a slope, therefore without critical impact of roll boresight error. Yaw estimation maybe biased by the residual roll and pitch errors since it uses full swath data over a slope. It is actually the case in practice, the yaw boresight remains the most difficult to estimate, which is due to the fact the patch-test procedure uses biased data and makes inappropriate assumptions.

In summary, we have seen that

1. Each patch of non-planar surfaces is distorted by "local" rotations which depends on swath attitude angle and therefore on local grazing angles;
2. Boresight decoupling assumptions are not valid, since each boresight angle which has not yet been corrected may distort a non-planar surface.

2 Boresight Estimation Methods

As mentioned in [3], the elimination of the systematic errors from survey data can be done by two different approaches. The first consists in analyzing each component of a survey system (ranging system, inertial measurement unit, positioning system, acquisition software), and characterizing individual errors from all sensors.

Another approach is to identify systematic errors from geo-referenced data, which happens to be corrupted by coupled and non-linear combinations of sensors errors. These methods aim at retrieving systematic errors by inversion methods. Then, calibration methods fall into two main classes.

2.1 Surface Matching Methods

From several overlapping swaths, DTM surfaces are constructed, generally by using TINs (Triangulated Irregular Networks). The goal is then to find the boresight rotation matrix corresponding to the best fit of the two surfaces. Several surface matching algorithms have been proposed, see [2, 4, 7]. Examples are the Iterative Closest Point, or normal matching methods. The idea behind normal matching is to define from a DTM an orientation vector (the normal). From one surface to another (e.g., for two overlapping swaths from two points of view) alterations of the normal vectors are the basis for calibration parameter estimation. The estimation process begins based on an iterative least squares method.

2.2 Tie Point Methods

This class of methods [6, 8, 10] consists in adjusting the calibration parameters from a limited data set containing targets or control points. The drawback of these methods is that they require the *a priori* knowledge of target points, which is feasible for land survey application, but obviously not for marine survey ones. One type of such methods does not require the knowledge of geolocalized target points, but requires to be able to determine a representative position of the target (center of a sphere, for instance) from ranging data. This kind of method is employed in Terrestrial Laser Scanning applications [5], where static scans are possible, and the scanning resolution is so high that the center of a sphere can be fitted with high accuracy. This class of methods could be transposed to MBES calibration, but would impose the design of specific targets, and a radical change in MBES calibration procedures.

3 Automatic MBES-IMU Boresight Calibration

The methods we propose are based on both classes of methods presented before and the following points:

- The use of a spatial reference model taking into account boresight angles, lever-arms and other source data provided by the survey sensor suite (positioning, IMU, MBES);

- The definition of an observation equation expressing the fact that overlapping data should coincide;
- The definition of an automatic data selection process which returns appropriate overlapping subsets;
- Adjustment methods which provide numerical estimation of the boresight angles;
- Statistical analysis tools that provide external and internal reliability of the estimation process, and returns boresight angle precision.

It has been mentioned that a global surface distortion due to boresight cannot be represented by a simple geometrical transformation like a similarity transformation for example. Indeed, from Fig. 2, one can readily see that both assumed (i.e. distorted) profiles (in green and blue) cannot be transformed into the actual profile represented in red. This simple observation enables us to classify several types of boresight calibration and estimation methods:

- *Rigorous*, methods and estimation procedures which estimate the boresight coordinate transformation from elementary sounding (e.g., points) or a subset of sounding from the same swaths. Indeed, these objects are submitted to a coordinate transformation which belongs to the class of transformations we are looking for.
- *Semi-Rigorous*, methods that estimate the boresight coordinate transformation using local overlapping surfaces patches.
- *Non-Rigorous*, all other methods.

We shall say that a boresight calibration method is a decoupling method if it ignores the coupling between roll, pitch and yaw. From this classification, we can say for example that the classical patch-test is a rigorous decoupling method. Referring to normal fitting methods, widely used in LiDAR applications, we can say that they are actually semi-rigorous, but non-decoupling: Indeed, these methods estimate normal vectors to local surfaces patches constructed from overlapping data sets (i.e., they are semi-rigorous) and they adjust in 3D the boresight angles in order to fit these normal vectors (i.e., they are non-decoupling methods).

3.1 Working Limits

Our aim is to design a 3D rigorous method, which can be easily automated by analyzing relevant overlapping swath data, and which provides boresight angle precision estimation. We present here a method which seems promising from preliminary experimental results.

Let us suppose that the boresight calibration data subset is a set of overlapping swaths, over a given area. We mention here that this area needs to be defined in a sense that all boresight angles will produce significant sounding errors, in other words all boresight angles should be observable. One should avoid for instance flat areas (for which pitch and yaw are not observable) and prefer slopes. One should also

Fig. 3 Fake boresight error
from overlapping data over
an edge, due to different
point of view and space
sampling effect

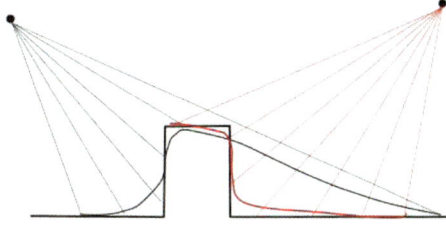

avoid areas containing edges (like wrecks for instance), since the sampling effect
between overlapping datasets will induce systematic boresight errors (see Fig. 3, for
which we cannot distinguish a DTM error due to the sampling effect from a boresight
error.

From the spatial referencing Eq. (1), assuming that latency is corrected (e.g.
known from either a systemic analysis or estimated), we have the Eq. (2).

$$\mathbf{X}_n(t) = \mathbf{P}_n(t) + C_{bI}^n(t)[C_{bS}^{bI}\mathbf{r}_{bS}(t) + \mathbf{a}_{bI}] \qquad (2)$$

For the sake of simplicity, we suppose that IMU data $C_{bI}^n(t)$ are not biased (i.e.,
the IMU is properly aligned with the local geodetic frame) and that MBES returns
are not subject to launch angle and range bias. This is actually the case whenever
the IMU is properly calibrated and aligned, sound speed profiles are known without
uncertainty, and the surface sound velocity is correctly measured and fed into the
MBES.

The parameters to be estimated are C_{bS}^{bI}, which depends on boresight angles
$(\delta\varphi, \delta\theta, \delta\psi)$, and (a_x, a_y, a_z), the three entries of the lever-arm vector a_{bI}.

Let us consider a cell from a grid defined over overlapping swaths. Within every
cell, we express the fact that if all points, corrected with appropriate boresight and
lever-arm values lie on a given quadratic surface, then the boresight and lever-arm
errors should be zero (see Fig. 4 below). From a practical point of view, if the grazing
angles of the MBES swaths cover a sufficiently wide interval (i.e., if the calibration
lines are run over a slope from distinct points of views), we should be able to estimate
the boresight angles. In other words, the boresight angles should be observable.

3.2 Observation Equation

We detail now how this problem can be expressed as an iterative least squares problem, and how the sounding uncertainties can be propagated through this least squares
problem in order to get estimates of both boresight and lever-arm precision.

Let us denote by **p**, the vector of (unknown) parameters defining a quadratic surface $S(\mathbf{p}; x, y, z) = 0$, and by χ, the vector of unknown boresight angles and lever-arm

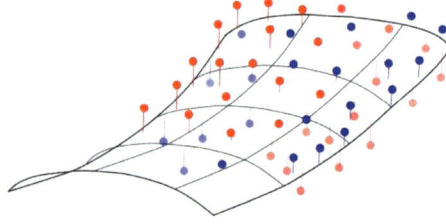

Fig. 4 Before boresight calibration, soundings from two overlapping swaths may not fit on a quadratic surface. In this example, the two point clouds do not match with any quadratic surface. In our approach, the quadratic surface and the boresight angles are adjusted in order to fit the overlapping point clouds

components. $\mathbf{p}$ can be chosen to be a 6 dimensional vector, and χ is a 6 dimensional vector. Using this notation we can write the Eq. (3).

$$\mathbf{X}_n = f(\chi; \mathbf{P}_n(t), C_{bl}^n(t), \mathbf{r}_{bS}(t)) \tag{3}$$

where $\mathbf{P}_n(t), C_{bl}^n(t), \mathbf{r}_{bS}(t)$ are here considered as external data depending on each sounding measured at time t. The criterion we use to determine both $\mathbf{p}$ and χ is expressed in the Eq. (4).

$$S(\mathbf{p}; f(\chi; \mathbf{P}_n(t), C_{bl}^n(t), \mathbf{r}_{bS}(t))) = 0 \tag{4}$$

Equation (4) express the fact that the point $\mathbf{X}_n(t)$ lies on a given quadratic surface. Let us now consider the collection of conditions, for all overlapping points of a given grid, defined on the horizontal plane. After linearization, this system, can be written as a least squares problem that can be solved by an iterative procedure, and enables both external and internal reliability analysis.

4   Numericals Results

We present some results, obtained from the application of the method presented above from calibration lines performed with an hydrographic system composed of an R2SONIC 2022, an IXBLUE OCTANS4, and a MAGELLAN proflex500 GNSS receiver. The data acquisition software used was QINSy. These tests have been conducted by the ENSTA Bretagne hydrographic team over a slope located in the Brest harbor.

Let us first mention that the geometry of line and overlaps used by our method is different from the classical patch-test method. Indeed, we need to guarantee boresight angle observability, which can be achieved only with a set of swaths obtained from significantly different points of view of the same area. Therefore, a set of cross-

Table 1 Calibration numerical results in the Brest harbor

Boresight angles [°]	Roll ($\delta\varphi$)	STD	Pitch ($\delta\theta$)	STD	Yaw ($\delta\psi$)	STD
Patch-test	0.62	?	1.64	?	1.88	?
ABE	0.679	0.006	1.657	0.002	1.995	0.03

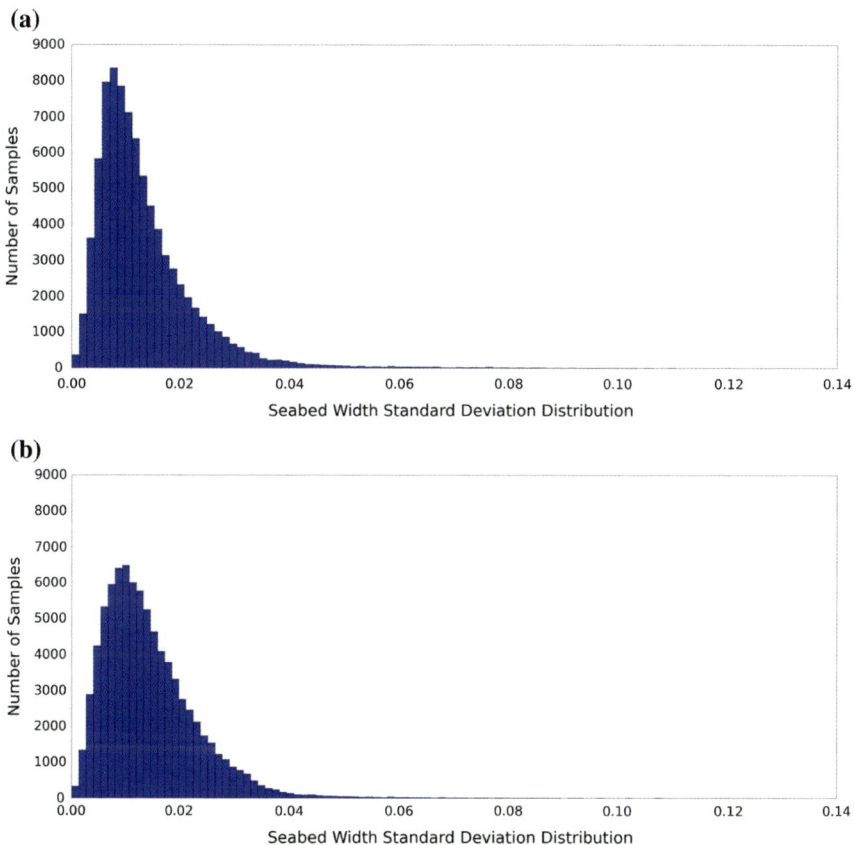

Fig. 5 From the two histograms, one can see that the automatic boresight method (**a**) provides a better global fit of overlapping data. Indeed the plot shows number or samples versus the adjustment error. **a** seabed width standard deviation histogram with estimated boresight angles correction. **b** seabed width standard deviation histogram with Patch-test boresight angles correction

ing lines over a slope has been surveyed. In order to compare our approach with the patch-test, we also performed patch test lines (over flat surfaces for roll, and the same slope for pitch and heading), and estimated calibration parameters with classical software tools. Table 1 presents the values of the boresight angles found by Automatic Boresight Estimation (ABE) and the classical patch-test.

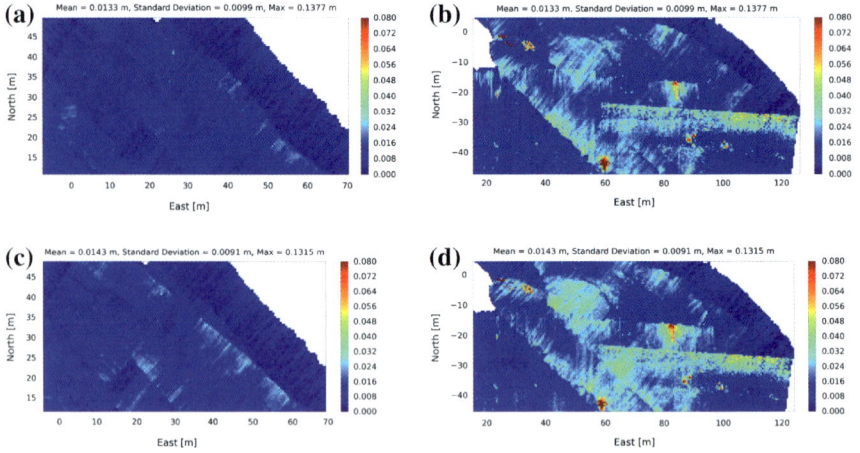

Fig. 6 Global data set precision estimation, for the boresight estimated by our automatic method (*top* with **a** and **b**) and a classical patch test (*bottom* with **c** and **d**). By observing the two selected areas, one can see that the automatic method is performing better. **a** seabed width standard deviation with the estimated boresight angles correction on a flat area. **b** seabed width standard deviation with the estimated boresight angles correction on a mix flat/slope area. **c** seabed width standard deviation with the Patch-test boresight angles correction on a flat area. **d** seabed width standard deviation with the Patch-test boresight angles correction on a mix flat/slope area

As a measure of the precision of the bathymetric surface built with a given boresight value, we use the following process: For each cell of a grid, we fit a plane by total least square (TLS) and use the orthogonal error of the point cloud which is given by the lowest singular value computed by the TLS. The advantage of this method with respect to the classical standard deviation map is to cancel out the effect of local slope.

Figure 5 shows the histogram of the orthogonal error (i.e., seabed width standard deviation) for both our approach and a classical patch test. Figure 6 presents the chart results obtained using the calibration results of the Table 1.

It is to be mentioned that the proposed approach is in theory able to estimate both boresight angles and lever-arms values through the same optimization process. However, from our preliminary results, it seems that the joint estimation of all these parameters is difficult from a practical point of view. Indeed, in order to obtain the observability of the boresight angle, we need significantly different points of view of a given smooth slope. In order to get lever-arm observability, we need relatively high attitude angles, again over a slope. It appears that our dataset contains only relatively high attitude angles over a moderate slope, and different overlapping lines over a sharp slope, but only with small attitude angle.

As a consequence, we see that we cannot systematically estimate both the boresight angle and the lever-arms from the same overlapping raw data, as observability of these parameters depends on source data. Therefore, the methodology we propose for the practical use of this approach is:

1. Estimate lever-arms from a data set selected using a lever-arm observability criterion;
2. Estimate the boresight angles from another area, selected using a boresight observability criterion.

5 Conclusion

The new calibration procedure introduced in this paper provides promising preliminary results. The use of an observation equation and least-squares optimization method allow working on the boresight problem source. Furthermore, the linearization of an observation equation gives us the possibility to use the statistical analysis toolbox of least-squares. All these aspects give the essential information (value, precision, internal and external reliability...) which should be in a calibration report. Moreover, this procedure, being automatic, allows the hydrographer to save time at sea.

The results presented need to be confirmed by other tests with different systems and other survey areas. The boresight angles estimation is reliable but the lever-arms estimation needs to be investigated in order to be included in a new global procedure.

References

1. Calder B. R., Brennan R. T, Malzone C., Marcus J., & Canter P. (2007). Application of High-Precision Timing to Distributed Survey Systems, *Proceedings of the US-Hydro Conference, Norfolk, VA.*
2. Filin, S. (2003). Recovery of systematic biases in laser altimetry data using natural surfaces. *Photogrammetric Engineering and Remote Sensing, 69,* 1235–1242.
3. Filin S., & Vosselman G. (2004). Adjustment of airborne laser altimetry strips. In: *ISPRS Congress Istanbul, Proceedings of Commission III.*
4. Glennie, C. (2007). Rigorous 3d error analysis of kinematic scanning lidar systems. *Journal of Applied Geodesy, 1,* 147–157.
5. Grejner-Brzezinska, D. A., Toth, C. K., Sun, H., Wang, X., & Rizos, C. (2011). A robust solution to high-accuracy geolocation: Quadruple integration of gps, imu, pseudolite, and terrestrial laser scanner. *IEEE Transactions on instrumentation and measurement, 11,* 3694–3708.
6. Kumari, P., Carter, W. E., & Shrestha, R. L. (2011). Adjustment of systematic errors in als data through surface matching. *Advances in Space Research, 47,* 1851–1864.
7. Morin K., & El-Sheimy, N. (2002). Post-mission adjustment methods of airborne laser scanning data. In: *FIG XXII International Congress, Washington DC,* 19–26 Apr 2002.
8. Schenk, T. (2001). Modeling and analyzing systematic errors of airborne laser scanners. Technical report, Department of Civil and Environmental Engineering and Geodetic Science, The Ohio State University, Columbus, OH.
9. Seube, N., Picard, A., & Rondeau, M. (2012). A simple method to recover the latency time of tactical grade IMU systems. *ISPRS Journal of Photogrammetry and Remote Sensing, 74*(2012), 85–89.
10. Skaloud, J., & Litchi, D. (2006). Rigorous approach to boresight self-calibration in airborne laser scanning. *ISPRS Journal of Photogrammetry and remote Sensing, 61,* 47–59.

New Tools for Seabed Monitoring Using Multi-sensors Data Fusion

C. Noel, C. Viala, S. Marchetti, E. Bauer and J.M. Temmos

Abstract SEMANTIC TS, an acoustics & oceanography engineering consulting company, uses sound to infer aquatic environment: water column, vegetation, bottom (nature and topography), sub bottom. SEMANTIC TS is able to operate, simultaneously or not, one or several acoustic systems, from very light survey units and to develop its own software suite devoted to data acquisition, processing, fusion and operational map production.

Keywords Bottom monitoring · Seabed · Acoustics · Side scan sonar · Multi-beam bathymetry · Multi-sensors data fusion

1 Introduction

Setting up a process to build accurate seabed maps is indeed challenging, but to ensure its reproducibility is even more complex. Now, this is a required condition since only evolution between two maps can provide us with relevant information to qualify occurring changes for monitoring purpose.

This paper details the methods used by Semantic TS since 2001 to set up affordable seabed monitoring techniques using multi-sensors data fusion. Costs of such methods are now affordable for both military and civil organisms, the latter facing an increasing number of norms about environmental monitoring.

This approach is aimed at designing light survey units to merge data from various sensors, at different frequencies and to set up an acoustics classification method for seabed nature using a dedicated software framework.

C. Noel (✉) · C. Viala · S. Marchetti · E. Bauer · J.M. Temmos
SEMANTIC TS Acoustics & Oceanography Engineering,
Sanary-sur-Mer, France
e-mail: noel@semantic-ts.fr

© Springer International Publishing Switzerland 2016   25
B. Zerr et al. (eds.), *Quantitative Monitoring of the Underwater Environment*,
Ocean Engineering & Oceanography 6, DOI 10.1007/978-3-319-32107-3_3

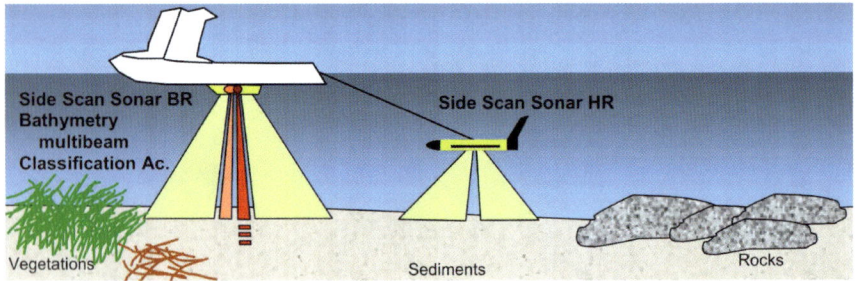

Fig. 1 Principle of multi-sensors data acquisition from SEMANTIC oceanographic survey vessel

2 Principle of Shallow Water Bottom Monitoring

To enable us to build maps and to monitor shallow water seabeds and vegetation, we propose to use the following tools:

- Small dedicated oceanographic survey ship
- Side scan sonar and hull interferometer to get multi-beam and Side scan imagery at the same time
- Detection and monitoring methods, Vertical Acoustic monitoring, CLASS (CLassification des Sédiments Superficiels) and FISH (Halieutique) improved since 2003 and using signals from a scientific echo sounder (SIMRAD ES60)
- Additional sensors: acoustic camera, sediments sounder, magnetometer,…

These systems, working at different frequencies, provide us with complementary information about the marine medium (Fig. 1).

3 Development of New Tools Devoted to Bottom Monitoring

3.1 Light Survey Units

We have developed small new survey units to be deployed in littoral sea areas, rivers, ponds, lakes.

Small size survey units offer high level of technology, both for platform positioning systems and for acoustic sensors. Boats are equipped with motion central and high speed internet is available through 3G used for D-GPS RTK corrections, from land reference D-GPS station, in real time. Survey units are able to produce energy to process simultaneously all the instrumentation (24/7 for the SEMANTIC unit) (Fig. 2).

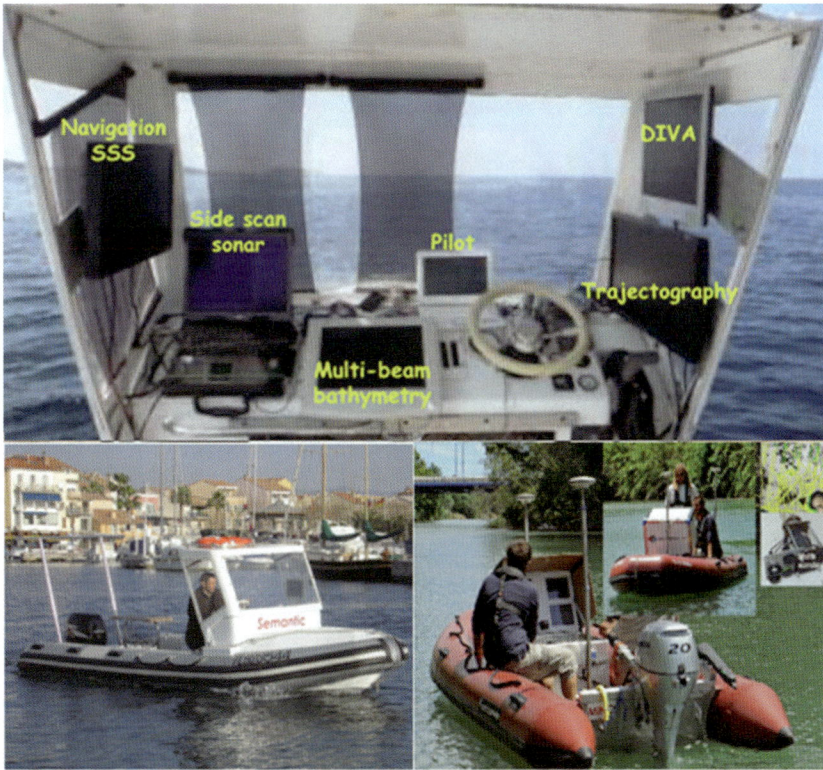

Fig. 2 SEMANTIC (6.5 m) and MINO (4.3 m) mini oceanographic survey vessels

3.2 Implementing Seabed Acoustics Classification Methods

Since 2004, we conduct studies on acoustics classification of vegetation and underwater sediments, which led us to release an innovative automatic seabed acoustics classification system (SACLAF), inferring the reflected signal in water column and seabed named SIVA (Système d'Inspection Verticale Acoustique).

3.3 Implementing Multi-sensors Data Fusion

We are working on acoustic data fusion from the following sensors since 2007:

- 3D Bathymetry (underwater topography of the location)
- Bathymetric roughness, providing information about the vegetation
- Side Scan Sonar imagery, where gray level gives information about bottom reflectivity and consequently on the vegetal (or non-vegetal) nature of the seabed

- Whether dense vegetation is present or not can be found out by the DIVA method [1]
- Sediment classification information provided by CLASS [2]
- Geo referencing of fishery resources provided by the FISH method

The uniqueness of our work lies in the various acoustics devices we can integrate simultaneously along with the specially developed software framework for the seabed monitoring.

This software acts as a scheduler, driving all devices and sensors, handling their synchronization, timestamping, conflicts, geo-referencing, the communication with the data acquisition station as well as the raw data storage.

This software also includes a scientific database featuring signal processing functions dedicated to acoustic classification, generating secondary data (process data: bathymetry, side scan sonar mosaic, results from SIVA classification method...).

These systems, working at different frequencies, provide us with complementary information about the marine medium. Data gathered from the various instruments is accurately georeferenced and time stamped (synchronized on the same time base) by the same DGPS RTK/Motion sensor (centimetric precision) positioning system. This common Space-Time reference basis, easing the data fusion process, significantly improves our knowledge of the marine medium and the performance and reliability of the monitoring process.

4   Results

Initially developed for Posidonia detection, the multi-sensor data fusion method is now regularly used in operation for various kinds of seagrass meadow... as well as bottom colonizing species: mussels, slipper limpet (*Crepidula Fornicata*)... [3].

These methods have been successfully applied in Corsica and French Riviera vegetation (on Mediterranean *Posidonia Oceanica and Cymodocea Nodosa* meadows and sediments), in Guyana (high turbidity), in French Brittany (*Laminaria Hyperborea*), in Arcachon basin (*Zostera Marina*). Following picture shows an example of data fusion results obtained near Sanary s/Mer (Riviera), on posidonia meadow (Figs. 3 and 4).

Fig. 3 Bathymetric micro-rugosity and Acoustic classification. (vegetation (*green*), fine sediment (*yellow*) et coarse (*orange*)) On side scan sonar mosaic and aerial view

Fig. 4 *Left* Isobaths. *Right* Corresponding side scan sonar imagery. *Black area* are full of cymodocea. *Dark gray areas* contain posidonia

5 Conclusions and Perspectives

Data fusion concept is innovative and powerful. It allows producing like in medical applications, very accurate 3D scan pictures of seabed derived from different sources (side-scan, multi-beams, echo sounder) and information (aerial pictures, classification methods results, divers/video observations ...). Power of data fusion concept remains on the quality of the data and on their complementarities. In this

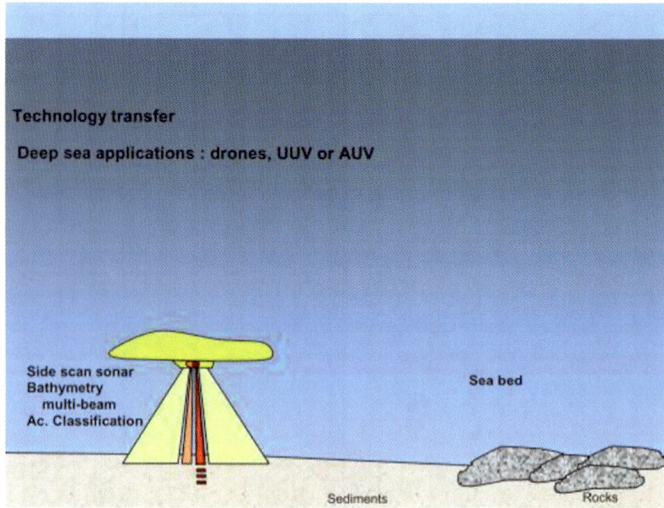

Fig. 5 Technology transfer to deep sea application

context such mini-survey units, able to operate and synchronize several complementary high resolution acoustic sensors simultaneously, and to precisely process
motion and geo-positioning, appears as a very efficient tool in the crucial data
collection first step of the data fusion process.

SEMANTIC TS is currently working on extending the SIVA method to acoustic
detection of coralligenous and to the characterization of posidonia dead matte.

Please note that the previously described methods, applicable for shallow water
monitoring, can be transposed to deep sea (embedded on drones), as illustrated in
the picture below. We are currently actively driving this technology transfer
(Fig. 5).

Acknowledgments The authors thank the D4S/MRIS from DGA as well as Agence de l'Eau
RMC which supports financially this work.

References

1. Viala, C., Noel, C., Coquet, M., Zerr, B., Lelong, P., & Bonnefont, J.-L. (2007). Pertinence de
   la méthode DIVA pour l'interprétation des mosaïques sonar latéral. In *Third Mediterranean
   Symposium on Marine Vegetation*, Marseille.
2. Chivers, R. C., Emerson, N., & Burns, D. R. (1990). New acoustic processing for underway
   surveying. *The Hydrographic Journal, 56*, 8–17.
3. Noel, C., Viala, C., Coquet, M., Marchetti, S., Bauer, E., Emery, E., et al. (2010). *Comparison
   of coastal marine vegetation mapping method*, Paralia Éditions.

The European Marine Data and Observation Network (EMODnet): Your Gateway to European Marine and Coastal Data

Jan-Bart Calewaert, Phil Weaver, Vikki Gunn, Patrick Gorringe and Antonio Novellino

Abstract Data from the marine environment are a valuable asset for marine industries, decision-making bodies and scientific research, but in Europe marine data are stored in a wide range of national, regional and international databases and repositories using different formats and standards which makes it difficult to find, assemble and use them efficiently. The European Marine Observation and Data Network (EMODnet) is a network of organisations set up in 2007 by the European Commission in the framework of EU's Integrated Maritime policy to address the fragmented marine data collection, storage and access in Europe. This paper introduces EMODnet in the context of EU's Marine Knowledge 2020 Strategy and highlights some of the main features of the EMODnet Data Portals, as well as those being developed by the EMODnet central portal. Finally, the paper zooms in on the features and endeavours of one specific EMODnet activity: the EMODnet portal with access to near-real time and historical data of physical parameters.

Keywords EMODnet · Marine data management · Data portal · Open access · EMODnet physical parameters

J.-B. Calewaert (✉)
EMODnet Secretariat, Ostend, Belgium
e-mail: janbart.calewaert@emodnet.eu

P. Weaver · V. Gunn
Seascape Consultants, Romsey, UK
e-mail: phil.weaver@emodnet.eu

V. Gunn
e-mail: vikki.Gunn@emodnet.eu

P. Gorringe
EuroGOOS AISBL, Brussels, Belgium
e-mail: patrick.gorringe@eurogoos.eu

A. Novellino
ETT SpA, Genoa, Italy
e-mail: antonio.novellino@ettsolutions.com

© Springer International Publishing Switzerland 2016   31
B. Zerr et al. (eds.), *Quantitative Monitoring of the Underwater Environment*,
Ocean Engineering & Oceanography 6, DOI 10.1007/978-3-319-32107-3_4

1 Introduction

Data from the marine environment are a valuable asset. Rapid access to reliable and accurate information is vital to obtain the knowledge necessary to address threats to the marine and coastal environment, in the development of policies and legislation to protect vulnerable areas of our coasts and oceans, in understanding trends and in forecasting future changes. Likewise, better quality and more easily accessible data is a prerequisite for innovation and further sustainable maritime economic development or 'Blue Growth'.

The costs of acquiring marine data through ocean observations in the EU is enormous, estimated at 400 million euro per year for data from remote sensing using satellites and more than 1 billion euro per year for collecting in situ data by public authorities [1]. In Europe, these costs are largely carried by the Member States. In addition, private bodies spent about 3 billion euro annually on sea and ocean data gathering and monitoring [1].

While access to marine data is critical for marine industries, decision-making bodies and scientific research, up to now it has been difficult to find, access, assemble and apply the data collected through observations in Europe. This is because most of Europe's marine data resources are collected by various local, national and regional entities and stored in unconnected databases and repositories. When available, the data are often not compatible making aggregation and wider scale use impossible. Recent studies have revealed that making high quality marine data held by EU public bodies more widely available would improve offshore operators' efficiency and save about 1 billion euros per year in gathering and processing marine data for operational and planning purposes. It would also stimulate competition and innovation in established and emerging maritime sectors, estimated at 200–300 million euro per year [2]. In addition it would improve efficiency of marine planning and legislation and reduce uncertainty in our knowledge and ability to forecast the behaviour of the sea.

To address the fragmented marine data collection, storage and access in Europe, the European Commission initiated the development of the European Marine Observation and Data Network (EMODnet) in the framework of EU's Integrated Maritime policy in 2007. The primary aim of EMODnet is to unlock existing but fragmented and hidden marine data and make them accessible for a wide range of users including private bodies, public authorities and researchers. At the onset of 2015, EMODnet consists of more than 160 organisations working together to observe the sea, to make the marine data collected freely available and interoperable, to create seamless data layers across sea-basins and to distribute the data and data products through the internet.

2 The EMODnet Development Process

The term EMODnet was first coined in 2006 in the preparations of the EC Integrated Maritime Policy as a way to provide a sustainable focus for improving systematic observations (in situ and from space), interoperability and increasing access to data, based on robust, open and generic ICT solutions [3]. The aim has always been to increase productivity in all tasks involving marine data gathering and management, to promote innovation and to reduce uncertainty about the behaviour of the sea. EMODnet has since been promoted as a key tool to lessen the risks associated with private and public investments in the blue economy, and facilitate more effective protection of the marine environment.

Since its adoption as a long-term marine data initiative, EMODnet has been developed through a stepwise approach in three major phases.

- Phase I (2009–2013) developed a prototype (so called ur-EMODnet) with coverage of a limited selection of sea-basins, parameters and data products at low resolution;
- Phase II (2013–2016) works towards an operational service with full coverage of all European sea-basins, a wider selection of parameters and medium resolution data products;
- Phase III (2015–2020) will work towards providing a seamless multi-resolution digital map of the entire seabed of European waters providing highest resolution possible in areas that have been surveyed, including topography, geology, habitats and ecosystems; accompanied by timely information on physical, chemical and biological state of the overlying water column as well as oceanographic forecasts.

Currently EMODnet is in the 2nd phase of development and provides access to marine data, metadata and data products spanning seven broad disciplinary themes: bathymetry, geology, physics, chemistry, biology, seafloor habitats and human activities. These data are being used to create medium-resolution maps of all Europe's seas and oceans spanning all seven disciplinary themes—these are expected to be complete in 2015. The next phase of EMODnet will involve the development of multi-resolution sea basin maps, commencing in 2015.

The development of EMODnet is a dynamic process so new data, products and functionality are added regularly while portals are continuously improved to make the service more fit for purpose and user friendly with the help of users and stakeholders.

Each theme is looked after by a partnership of organisations that have the necessary expertise to standardise the presentation of data and create data products. From the onset, EMODnet has been developed based on a set of core principles:

- Collect data once and use them many times;
- Develop data standards across disciplines as well as within them;
- Process and validate data at different scales: regional, basin and pan-European;

- Build on existing efforts where data communities have already organised themselves;
- Put the user first when developing priorities and taking decisions;
- Provide statements on data ownership, accuracy and precision;
- Sustainable funding at a European level to maximise benefit from the efforts of individual Member States;
- Free and unrestricted access to data and data products.

3 Overview of EMODnet Thematic Data Portals

3.1 Introduction

For each of its core themes, EMODnet has created a gateway to a range of data archives managed by local, national, regional and international organisations. Through these gateways, users have access to standardised observations, data quality indicators and processed data products, such as basin-scale maps. These data products are free to access and use.

3.2 EMODnet Bathymetry

Bathymetry is the information that describes the topography of the seabed, as depth from the sea surface to the seafloor. It is an essential component in understanding the dynamics of the marine environment: the shape of the seabed is controlled by the underlying geology, and it exerts a strong influence on ocean circulation and currents, local fauna and seafloor habitats. Safe ocean navigation relies on accurate bathymetry data, which are also essential for planning marine installations and infrastructure such as wind turbines, coastal defences, oil platforms and pipelines. Bathymetry forms the foundation of any comprehensive marine dataset; without it, the picture is incomplete.

Currently, EMODnet provides bathymetric data and data products for all European sea basins. Users have access to Geographical Information System (GIS) layers covering water depth on a grid of up to 1/8 min latitude and longitude or in vector form at 1:100 000 scale; depth profiles along survey tracks, and survey metadata.

Users can download digital terrain model (DTM) data products that can be used in combination with other data layers from within EMODnet. A continually-updated data discovery and access service allows users to identify and request access to the underlying bathymetric survey data, held by a range of organisations, which form the basis of the DTM products. As EMODnet evolves, current bathymetry maps will be regularly updated with new data and complemented by coastal maps where the resolution is as high as the underlying data allows.

3.3 EMODnet Geology

Geological data are collected in a number of ways: physical samples via coring, drilling, grab sampling or dredging; direct observations using towed cameras and remotely operated vehicles, and acoustic remote sensing techniques that give an indication of the seafloor substrate. Of these techniques, only drilling or coring can reveal more than just the surficial geology. To probe deeper into the sub-seafloor, seismic survey methods are required.

Primary geological survey information requires significant expert interpretation to generate maps, and geological data are often used in combination with bathymetry to build up a comprehensive picture of the seabed. These data are a vital component of seafloor habitat maps, and are essential tools in marine spatial planning, coastline protection, offshore installation design, environmental conservation, risk management and resource mapping.

Currently, EMODnet provides access to geological data and maps at a resolution of 1:250 000 wherever possible that provide information on seabed substrate, seafloor geology (including boundaries, faults, lithology and age), sediment accumulation rates, coastline erosion and migration, areas of mineral resources, and the location and probable frequency of significant geological events such as earthquakes and volcanic eruptions.

3.4 EMODnet Seabed Habitats

Habitat maps are constructed from a number of basic data layers containing physical data that describe the environment in any given location. A habitat type is then derived on the basis of those environmental characteristics. EMODnet uses the latest European Nature Information System (EUNIS) habitat classification, which is the standard system in operation across Europe.

EMODnet provides a predictive seabed habitat map covering all European seas at 1:250 000 scale resolution. It builds on the broad-scale seabed habitat map developed under the EUSeaMap project, with enhanced validation and inclusion of regional and local habitat maps produced by Member States. EMODnet data on seabed substrate, energy at the seabed, biological zone and salinity at the seabed are combined to produce EUNIS habitat maps, with confidence indices to demonstrate the level of certainty for any given location.

Future developments will include increased emphasis on ecologically crucial coastal areas, with a view to mapping shallow inshore waters and eventually creating a seamless land-to-sea habitat map. Biologically-defined habitats, such as *Posidonia* seagrass beds in the Mediterranean will be included alongside habitats defined purely on their physical characteristics.

3.5 EMODnet Chemistry

Early detection, tracking and prediction of the movement of pollutants at sea are vital for the effective mitigation of their impacts on marine habitats and human infrastructure. Seawater chemistry data is used in combination with physical oceanographic data and bathymetry to trace the source of pollution, track its likely future trajectory, concentration and persistence, and to formulate a course of action to prevent or reduce impacts on the marine environment and to human well-being. Water chemistry data acquisition is often at the centre of routine monitoring efforts of Member States in response to national and European legislation or regional obligations.

EMODnet provides access to individual measurements as well as a range of products such as interpolated maps of chemical variables per region over time and graphics of station time series. Data include measurements of fertilisers, dissolved gases, chlorophyll, silicates, pH, Organic matter, synthetic compounds, heavy metals, hydrocarbons, radionuclides, and plastics.

3.6 EMODnet Biology

Measuring or observing marine life on a large scale is difficult. For the most part, data are collected over short time periods or in relation to specific species in target locations. Often, data are collected using different standards, technologies and conventions, making it challenging to combine information from different surveys or different databases.

EMODnet assembles these individual datasets and processes them into inter-operable data products for assessing the environmental state of ecosystems and sea basins. These data products illustrate the temporal and geographic variability of occurrences and abundances of marine phytoplankton, zooplankton, macro-algae, angiosperms, fish, reptile, benthos, bird and sea mammal species—in particular, introduced or harmful species, species of conservation concern and those used as ecological indicators.

Products include gridded map layers showing the average abundance of at least three species per species group for different time windows (seasonal, annual or multi-annual) using geospatial modelling and spatially distributed data products. Calculation of specific aggregated and gridded products indicating the presence, absence, abundance and diversity of species and communities can give an indication of ecosystem health and temporal trends for specific sea basins, which can be used to improve ecosystem-based management.

3.7 EMODnet Human Activities

Pressure on Europe's marine space and resources is at an all-time high. Continual demand for resources such as oil and gas, marine minerals and fish must be managed alongside the need to use marine space for renewable energy installations, communications cables, waste disposal sites and shipping. Additionally, societal demand for marine tourism and leisure activities, and the need to conserve marine ecosystems and habitats is leading to increasing competition and conflict between different marine sectors. Having access to accurate information to assist with planning, regulating and managing marine activities in a sustainable and responsible manner is critical.

EMODnet provides access to data describing the geographical position, spatial extent, and attributes of a wide array of human activities in the marine environment. From pipeline routes and waste disposal sites, to ports and protected areas, EMODnet maps activities or installations that could affect other ocean users, have an impact on the marine environment or that are themselves vulnerable to disturbance. It also provides a historical view of activities so that trends can be analysed and future requirements better anticipated.

EMODnet provides data and information on various human activities such as aggregate extraction, shipping (commercial/leisure), cultural heritage, dredging, fisheries zones, hydrocarbon extraction, major ports, mariculture, ocean energy facilities, pipelines and cables, protected areas, waste disposal (solids), wind farms, other forms of area management or designation.

3.8 EMODnet Physics

Europe's oceans and atmosphere are constantly measured and monitored through a network of remote, fixed and mobile in situ observing stations. The volume of data collected this way is substantial, ranging from the most fundamental information such as sea level, atmospheric pressure, sea temperature and salinity, to more complex measurements of turbidity and fluorescence in the water column. EMODnet provides a gateway to this vast resource of ocean physics data where users can access both near-real time (within a few hours of measurement) and historical archive data that are processed, validated and managed by oceanographic institutes and made available via EuroGOOS Regional Operational Oceanographic Systems (ROOSs) and National Oceanographic Data Centres (NODCs), combined with supplementary data from ongoing observing programmes such as EuroArgo. In what follows (Sect. 4), this paper specifically looks into the operation, functionality and data made available by EMODnet Physics as an example of a mature thematic portal illustrating how EMODnet helps to make marine data more easily available and useful at European scale.

4 Access to Physical Data, Products and Services by EMODnet Physics

4.1 Introduction

The EMODnet Physics project started in 2010 and one year later the first EMODnet Physics Portal[1] was launched with the aim to establish a single gateway to near real-time data and historical time series and datasets covering a wide range of physical conditions of the European sea-basins monitored both by fixed and mobile observation platforms such as moorings, drifters, gliders and ferryboxes. Data layers currently available via EMODnet are diverse and include salinity, temperature, currents, oxygen, fluorescence, pH, turbidity, sea level, wave height and period, horizontal wind speed and direction, atmospheric pressure, dew point, dry air temperature, humidity, light attenuation. Data on changes in sea-level and ice cover will be added shortly.

Rather than starting from scratch, the EMODnet Physics data portal builds on the efforts of two major existing European marine data infrastructures, adding high level services, features and functionalities but without duplicating services or adding complexity: (i) the Regional Ocean Observing Systems or ROOSs[2] which are the main components of the European Global Ocean Observing System (EuroGOOS) empowered via the MyOcean[3] project providing access to in situ near real-time data; and (ii) the network of National Oceanographic Data Centres (NODCs) organized under the SeaDataNet[4] (SDN) project providing access to delayed-mode data. EMODnet Physics brings together these groups and their infrastructures, mobilising considerable know-how in the collection, processing, and management of ocean and marine physical data as well as expertise in distributed data infrastructure development and operation at pan-European level.

4.2 EMODnet Data Categories

EMODnet Physics makes available near real time data and metadata provided by national data originators organized at EuroGOOS Regional level according the ROOSs infrastructure. It provides free and open access to all available near real time data of the latest 60 days to any user without the need for registration. Table 1 shows the number of operational platforms that provided at least one dataset for the past 60 days (from 16 February). Operational platforms provide data time series as

---

[1]http://www.emodnet-physics.eu/portal.

[2]http://www.eurogoos.eu.

[3]http://www.myocean.eu.

[4]http://www.seadatanet.org.

Table 1 Number of platforms connected to EMODnet Physics providing at least one dataset during the last 60 days (from 16 February 2015)

	Drifting buoys (DB)	Ferrybox (FB)	Gliders (GL)	Fixed buoys or mooring time series (MO)	Profiling floats vertical profiles (PF)	Argo floats (AR)	Total
Number of platforms	35	6	1	666	26	490	1220

soon as data is ready, for example a fixed platform is delivering data daily, an ARGO float is delivering almost weekly.

To access data older than 60 days (monthly archives) however, users need to register and log into access the system. Validated delayed data series and metadata are organized according to, and in collaboration with, the network of NODCs. By the end of 2014, EMODnet Physics fully integrated historical data datasets for 794 platforms (11,450 datasets) which are made available via the SeaDataNet Common Data Index (CDI) discovery and request system.

EMODnet Physics covers all European sea-basins but also provides data from measuring platforms beyond European waters. The number of measuring platforms connected to the system varies from sea-basin to sea-basin with the Black Sea lagging behind in terms of number of platforms and measured parameters (Table 2).

While users are able to download raw data from these platforms, a range of data products such as time series and plots are also provided for users according to the different data types. New products are developed regularly. The most recent addition is the creation of a product for platforms measuring wind, providing users with a set of plots showing among others average, maximum and minimum wind strengths over time for a given period (Fig. 1).

4.3 Access Provision Modalities and Services

The EMODnet Physics Portal consists of two main components: (i) a landing page with general information on the project approach, available data and products from connected platforms as well as the most important and widely adopted quality control procedures; and (ii) a set of services providing user interfaces and functionalities to search, visualise and retrieve data and products. The main services include (i) a map viewer to search, visualise and retrieve data and products; (ii) a catalogue which provides similar functionalities to the map viewer but in a more text based interface without visualising the platform locations on a map; and (iii) a range of interoperability services for machine-to-machine communication.

Table 2 Number of operational platforms connected to EMODnet Physics per sea region and per group of parameters

	Wave and winds	Temperature	Salinity	Currents	Light attenuation	Sea level	Atmospheric	Others	Chemical	Subtotals
Artic, Barents, Greenland, Norwegian Sea	2	48	43	2	0	37	3	52	12	199
Atlantic, Bay of Biscay, Celtic Sea	94	427	287	84	12	179	68	366	77	1594
Baltic Sea	3	11	7	5	0	47	1	7	1	82
Black Sea	0	1	1	0	0	1	0	1	0	4
Global Ocean	1	120	104	8	0	5	3	126	13	380
Mediterranean Sea	105	244	123	157	23	112	55	157	39	1015
North Sea	7	16	13	12	1	41	6	9	5	110
Other (e.g. Land platforms)	0	4	2	1	0	25	2	4	1	39
Total	**212**	**871**	**580**	**269**	**36**	**447**	**138**	**722**	**148**	**3423**

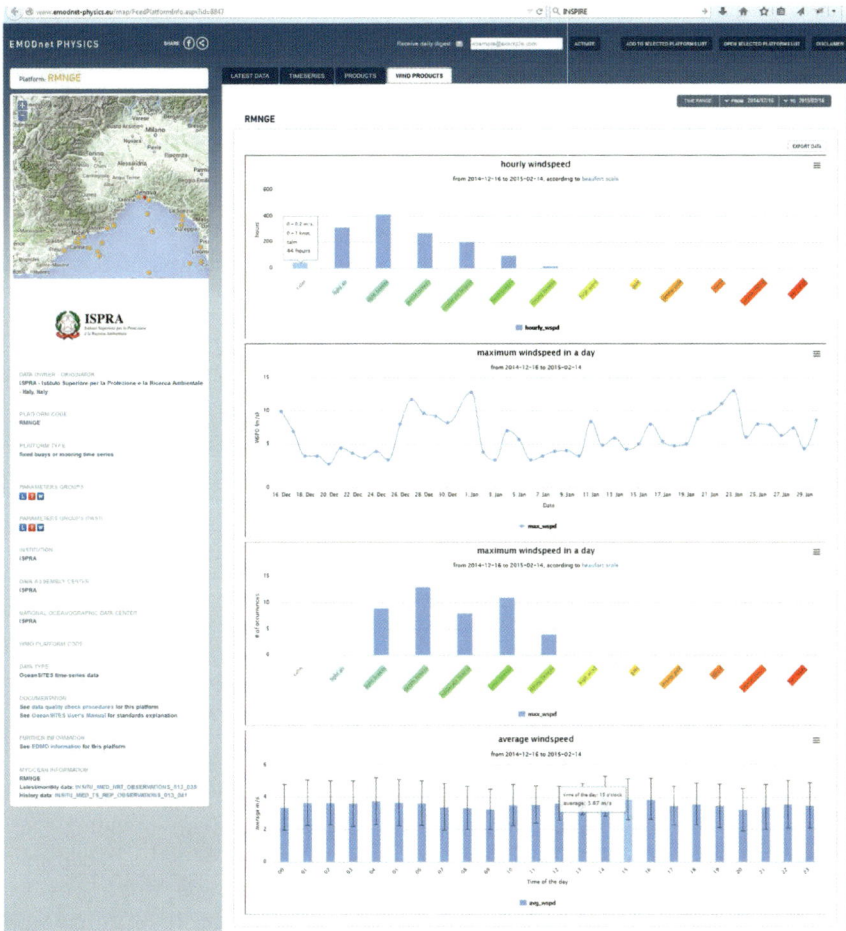

Fig. 1 EMODnet Physics overview of wind products for a specific platform operated by the Italian Institute for Environmental Protection and Research (ISPRA) in the Mediterranean

The EMODnet Physics Map Viewer

The EMODnet Physics map viewer[5] is designed to facilitate data discovery and selection (see Fig. 2). The map viewer visualises the location of platform connected to the system. For mobile platforms the latest position is shown. The viewer provides filters to allow users to identify and select a subset of available data based on the type of platform (e.g. fixed stations, drifting buoys, etc.), physical parameter (e.g. sea temperature, sea level, waves and winds), sea basin, country and data provider. By selecting a platform, a small panel summarizes what the platform is

---

[5]http://www.emodnet-physicis.eu/map.

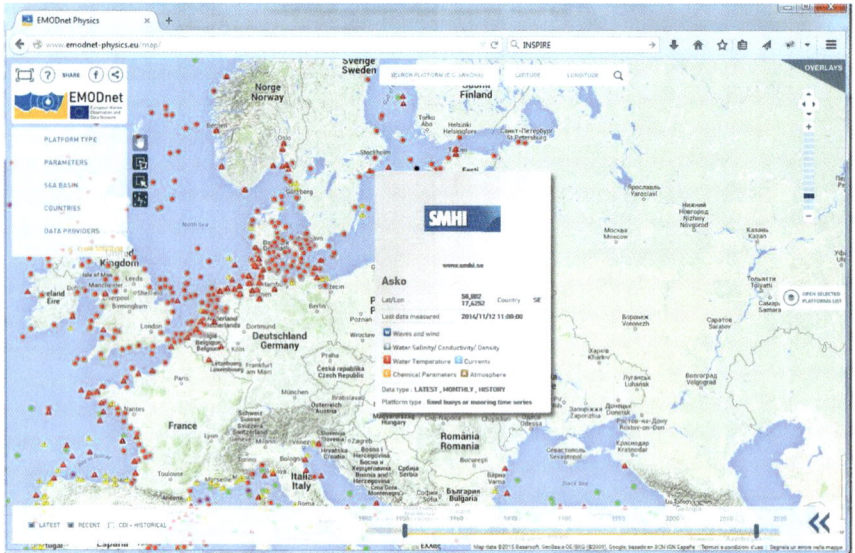

Fig. 2 EMODnet Physics map viewer showing a selected platform summary panel. Fixed stations are shown in *red*, drifting buoys in *violet*, gliders in *yellow*, profilers in *blue*, ARGO floats in *green* and ferryboxes in *black*

providing (parameters and latest measurements) as well as some basic platform information such as platform owner.

The map viewer also provides an option to select the "time range" allowing users to define a window in time to obtain an overview of the platforms providing data in the selected time period. There are basically three categories of data sets depending on the type of data providers: (i) 'latest data' refers to platforms which provide at least one dataset for the past 60 days; (ii) 'recent data' relates to all platforms which are still operational but not necessarily providing datasets in the last 60 days; while (iii) 'CDI-historical' refers to platforms with datasets validated by national oceanographic data centres.

But the map viewer does more than that: by clicking on a specific platform a dedicated platform page is provided which allows user to easily see available parameters, plot previews, see data availability, provided products (monthly averages, max and min products, and if available wind products) and available validated datasets (see Fig. 3). The platform page also provides a daily digest service which allows users to register for email updates on a daily base with the latest values of measured parameters for selected platforms.

Interoperability Services

EMODnet Physics provides a range of interoperability services for machine-to-machine communication for easier data integration and accessibility. Core services include the provision of web services, Web Map Service (WMS) and

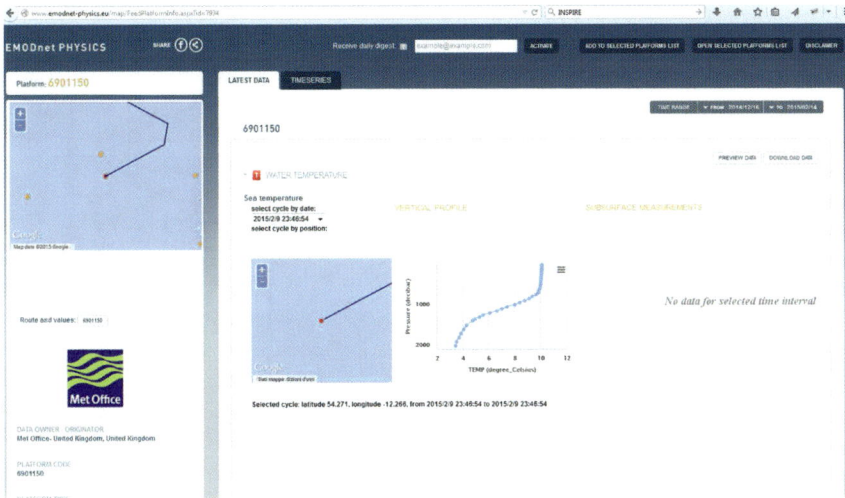

Fig. 3 EMODnet Physics platform view showing data plots for an ARGO float

Web Feature Service (WFS). EMODnet Physics is exposing fully OGC compliant WMS/WFS layers by exploiting a GeoServer[6] based infrastructure. By linking to these interoperability services many users are already adding EMODnet Physics data into their own applications and products. For example, these services allow an easy uptake of EMODnet Physics datasets into the Ocean Data Portal (ODP[7]) from the International Oceanographic Data and Information Exchange (IODE) programme of UNESCO's Intergovernmental Oceanographic Commission (IOC) giving the global visibility to the EMODnet physics data providers (see Fig. 4).

4.4 Concluding Remarks and Further Developments

Since its inception in 2011, the EMODnet Physics portal has attracted a steadily increasing number of visitors and users, greatly assisted by a dedicated communication strategy and a user-oriented approach. For example, the portal infrastructure automatically sends e-mails to more than 1000 users whenever a news item is published and users are updated about new datasets matching their interests when they are made available. As a result, by the end of 2014 each month the portal receives about 600 unique visitors and more than 20,000 physics data downloads are requested monthly (partly manual partly through machine to machine communication). By providing access to physical data and metadata, EMODnet Physics

---

[6]http://geoserver.org/.

[7]http://www.oceandataportal.net/portal/portal/odp-theme/data/nodcs.

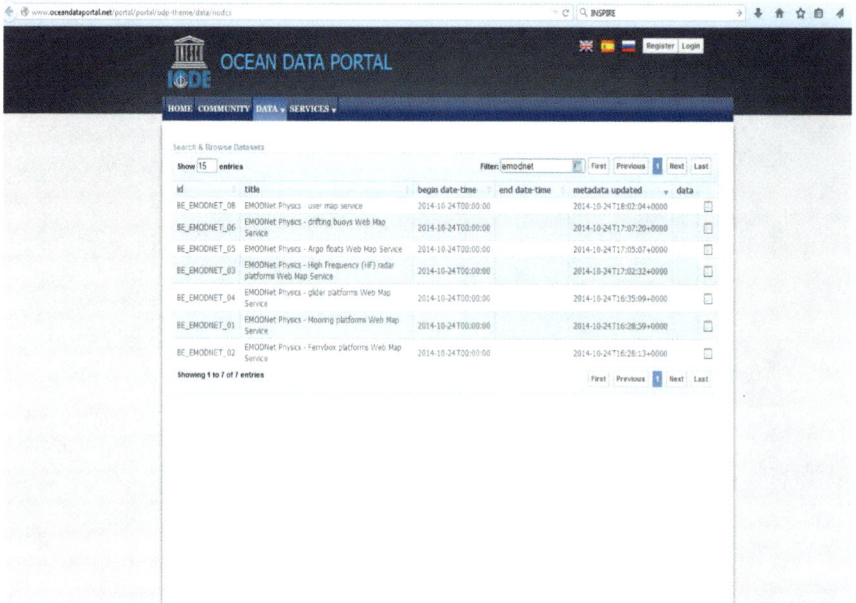

Fig. 4 External users such as the Ocean Data Portal (ODP) are connecting to the EMODnet Physics interoperability services to visualise EMODnet data layers

also contributes to important global initiatives such as the Ocean Data Portal (ODP) and the Global Monitoring for Environment and Security (GMES) marine core service.

EMODnet Physics is under continuous development to improve the streamlining and optimisation of the data flow, to gradually extend the provided features and strengthen the backoffice infrastructure to ensure data and information remains only one click away from the user. Future developments will include closer collaboration with the marine component of the EU's Copernicus earth observation programme and wider coverage of underrepresented sea-basins with a specific target to fill gaps in time series.

5   EMODnet Central Portal

To improve user experience and strengthen the coherence and functionality of EMODnet as a whole, a central 'EMODnet Entry Portal' has been established to provide an entry point (http://www.emodnet.eu) delivering access to data, metadata and data products held by EMODnet thematic sites as well as developing data products and search results combining data from several thematic portals. Since October 2014, the central portal made available a first data service called the EMODnet Central Portal Query Tool which allows user to simultaneously access data layers made available by the different EMODnet thematic portals, combining them in one single output. The tool will be gradually expanded with more parameters, search options and functions to allow manipulation of the outputs.

6 EMODnet Sea-Basin Checkpoints

User requirements are a priority in EMODnet, so a series of sea-basin 'Checkpoints' are being put in place, starting with the Mediterranean and North Sea in 2013. These regional mechanisms have been established to assess the observation capacity in all regional sea-basins from the perspective of concrete application areas (e.g. spill response, offshore installation siting, etc.). EMODnet Checkpoints are expected to identify whether the present observation infrastructure is the most effective possible, and whether it meets the needs of users. Tenders for additional regional assessment hubs covering the Arctic, Atlantic, Baltic and Black Sea have been launched in 2014 and are expected to be initiated early in 2015.

7 Coordination and Monitoring

Since September 2013, EMODnet activities are supported by a dedicated Secretariat responsible for public relations and communication, coordination of activities and monitoring of the performance of the EMODnet projects. Under supervision of DG MARE and the EMODnet Secretariat, a Steering Committee consisting of coordinators from EMODnet portals as well as observers from the European Environment Agency (EEA) and the European Commission Directorate-General for Environment (DG ENV) oversees the development of the Central EMODnet portal and guides the development of complementary services. This has greatly improved internal collaboration and exchange of best practices leading to a more coherent development of all EMODnet projects working towards a common goal.

8 Conclusions

EMODnet is a long-term marine data initiative supporting a sustainable blue economy in Europe, constructed through a stepwise approach. Halfway through its development, the resources and services are already useful and data portals are progressing rapidly to (i) become fully operational; (ii) provide the best available data, free of restrictions on use; and (iii) become more user friendly and fit for purpose. EMODnet Physics is a perfect example of how EMODnet, through the application of innovative technologies, can act as a gateway and one stop shop to access the vast resources of geo-referenced marine data in Europe supplemented with added value tools and functionality.

Bringing observations, products, services and knowledge to users and the public requires appropriate tools and guidance. Ensuring fitness for purpose can only be done together with a growing number of data providers and users and EMODnet will increasingly rely on the involvement of stakeholders to guide further developments.

Today, more than 110 organisations are involved in the EMODnet programme and new contributors are always welcome. EMODnet will continue to strengthen its collaboration with other marine knowledge providers, including fisheries, the marine component of the EU's Copernicus programme and the private sector, to create a common platform for marine data in Europe. This will include work both upstream and downstream to ensure more data is ingested to enter into the EMODnet system, as well as making sure that the data and products are easy to find, obtain and used.

Acknowledgments The European Marine Observation and Data Network (EMODnet) is financed by the European Union under Regulation (EU) No. 1255/2011 of the European Parliament and of the Council of 30 November 2011 establishing a Programme to support the further development of an Integrated Maritime Policy.

We are very grateful for the support and valuable contributions of all EMODnet partners and particularly acknowledge the work of EMODnet project coordinators Dick Schaap (bathymetry), Alan Stevenson (geology), Jacques Populus (seabed habitats), Alessandra Giorgetti (chemistry), Simon Claus (Biology), Antonio Novellino (Physics), Alvise Bragadin and Alessandro Pititto (human activities).

References

1. European Commission. (2010). European Marine Observation and Data Network. Impact Assessment. Com(2010) 461 sec(2010) 999.
2. European Commission. (2014). Marine Knowledge 2020 Roadmap. COM(2014) 254 final.
3. European Commission. (2006). Green Paper. Towards a future Maritime Policy for the Union: A European vision for the oceans and seas. SEC(2006) 689.

Surveying in Hostile and Non Accessible Areas with the Bathymetric HydroBall® Buoy

Mathieu Rondeau, Nicolas Seube and Julian Le Denuf

Abstract This paper describes the performance analysis of an autonomous drifting buoy equipped with a GNSS receiver, an inertial measurement unit and a single beam echosounder. The system is intended for surveying difficult access areas like high-flowing rivers, confined zones and ultra shallow waters, which are unreachable using a classical survey launches.

1 Introduction

In the framework of dams construction and exploitation, there is a need to map riverbeds in support to hydropower infrastructure construction and maintenance. White water areas often show a limited access and high flows and therefore cannot be surveyed with a classical hydrographic survey launch. In 2008, motivated by a demand from the company Hydro-Quebec, the CIDCO realized a technical review of the available systems for such surveying tasks, and concluded that the development of a new system should be undertook. This system, called HydroBall®, provides a low cost integrated solution for bathymetric data acquisition in hostile and non accessible areas. Its spherical design and robust shell casing encloses a single beam echosounder, a GNSS receiver, a MEMs IMU and a bluetooth communication link.

Compared to existing drifting buoys [1–5], the HydroBall® system is intended to achieve hydrographic survey with a level of precision which complies with international and industrial standards, as it will be shown in the next sections by a Total Propagated Uncertainty analysis.

M. Rondeau (✉) · N. Seube
CIDCO, Rimouski, Canada
e-mail: mathieu.rondeau@cidco.ca

N. Seube
e-mail: nicolas.seube@cidco.ca

J.L. Denuf
ENSTA Bretagne, Brest, France
e-mail: julian.le_deunf@ensta-bretagne.org

© Springer International Publishing Switzerland 2016   47
B. Zerr et al. (eds.), *Quantitative Monitoring of the Underwater Environment*,
Ocean Engineering & Oceanography 6, DOI 10.1007/978-3-319-32107-3_5

After some successful trials for riverbed surveys, the range of application rapidly grew to confined area surveys, standard SBES hydrographic surveys and shore profiling surveys. Indeed, due to the fact that this system is a fully integrated SBES survey system it can be deployed from any opportunity platform.

The first section presents the system in terms of hardware integration and processing software as well as several survey projects that have been conducted using HydroBall®. In the second section we present the a priori Total Propagation Uncertainty (TPU) analysis of the system. In section three, the results are compared to actual a posteriori TPU observations obtained from surveys data.

2   The HydroBall® System and Its Applications

The HydroBall® system integrates a SBES operating at 500 kHz, a dual frequency GNSS receiver, a MEMs IMU and a bluetooth communication link (see Fig. 1). The system is fully autonomous, thanks to a micro-controller which hosts a data acquisition and management system. The system has a minimum autonomy of 24 h.

On operation, once the GNSS receiver is able to deliver a position, all data from the other sensors (SBES, IMU) are time-tagged and saved in raw data files. As the HydroBall® integrates low-cost sensors unable to take in input any timing information, all data are time stamped upon reception by the micro-controller. The micro-controller's clock is regularly reset on the GNSS time, as provided by the GNSS receiver.

As the HydroBall® system is intended for an autonomous usage, it is very important to guarantee the data quality, as no operator can handle any problem occurring within the system, in the same way a qualified hydrographer would operate a classical SBES survey system. Data quality is analyzed in the next section, thanks to a objective comparison between an a priori TPU computation and a a posteriori TPU observation.

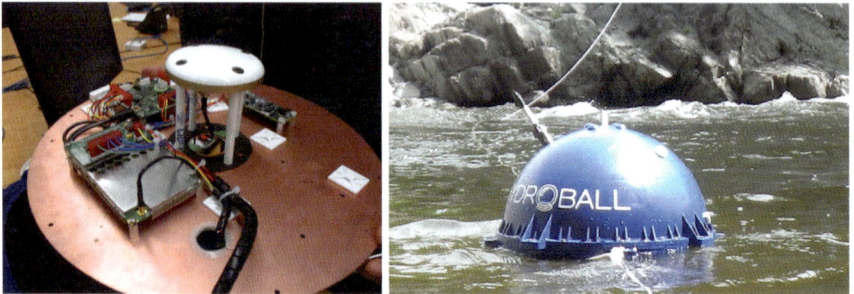

Fig. 1 The HydroBall® system is integrated in a 40 cm sphere. It is equipped with a SBES operating at 500 kHz, a L1/L2 GNSS receiver and a MEMs Inertial Measurement Unit

HydroBall® data processing is performed off-line and consists in three steps:

1. GNSS data post-processing: GNSS data are converted into RINEX format and the user can process these data in PPK mode, using corrections from a network of permanent station or from a fixed GNSS beacon. Note that the L1/L2 GNSS receiver can also compute position fixes in RTK mode;
2. Attitude and SBES returns are selected thanks to their time tag;
3. The computation of the corresponding sounding in the Local Geodetic Frame is performed: Post-processed GNSS data, attitude and SBES returns are merged by

Fig. 2 Some applications of the HydroBall® system: *Top left* Transect of a river (Rimouski river); *Top right* Riverbed survey (Rimouski river); *Middle left* Deployment from an inflatable (Anguille Lake); *Middle right* Deployment from an amphibious vehicle for beach profiling (Anse au Lard); *Bottom left* Survey in a confined area (Romaine river); *Bottom right* Deployment from an Helicopter for dangerous areas surveys (Romaine river)

a software written in Python which associates to any SBES return a sounding coordinated in the Local Geodetic Frame. This software implements appropriate corrections for latency and boresight angles between the SBES and the IMU.

In Fig. 2, we describe how the HydroBall® has been deployed for various types of surveys.

The primary usage of HydroBall® is riverbed surveys. Both transversal profiling surveys and longitudinal drifting surveys have been performed. It appeared that in practice, during survey project conducted by the CIDCO, HydroBall® was easier to deploy and set-up than a traditional pole mounting SBES survey system in the framework of classical single beam surveys. The main added-value of HydroBall® has been to enable us to survey non accessible areas where no traditional surveys means could be deployed:

- In ultra-shallow waters, HydroBall® exhibits good performances for projects that require both land survey and bathymetric survey data. For instance beach profiling is a typical application for which the HydroBall® compactness and full integration of GNSS and SBES are relevant. For this class of application, the system has been mounted on an Argo amphibious vehicle. In this configuration, the HydroBall® delivers SBES data until reaching the land (the SBES gives returns until a minimum depth of 10 cm) and is able to perform a mobile land GNSS survey while operating on the beach.
- In non accessible areas (canyons, kettles, etc.), HydroBall® can easily be deployed and recovered by hand.
- In areas where safety is an issue, HydroBall® has been deployed from an Helicopter and has shown to be an appropriate respond to challenging survey works. Indeed, the upstream section of one of the Romaine river rapid has been surveyed with this system.

3   Total Propagated Uncertainty of the HydroBall® System

The HydroBall® system can be described by:

- A reference point which is an arbitrary point of the HydroBall®. This point is the origin of all lever arms measurements.
- Lever arm (denoted by a_{bV} in Fig. 3), supposed to be measured in the (bV) frame, a frame defined in reference of the HydroBall® body itself.
- Frames attached to the SBES and the IMU. They are denoted respectively by (bS) and (bI).
- A local geodetic frame, or navigation frame used for platform orientation purposes.

Fig. 3 HydroBall® Sketch.
The lever-arm vector a_{bV} is
defined from the GNSS
antenna center of phase to
the SBES transducer
acoustic center

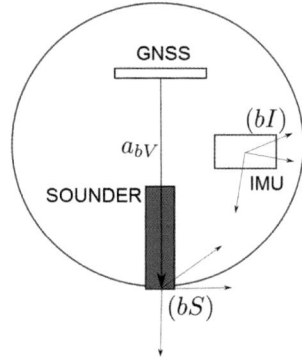

The single beam echo sounder returns will be denoted by

$$r_{bS} = \begin{pmatrix} 0 \\ 0 \\ \rho \end{pmatrix}$$

where ρ is the raw SBES return, supposed to be corrected from refraction due to the sound speed profile. We shall denote by C_{bS}^{bI} the boresight transformation matrix between the (bS) frame and the (bI) frame. This transformation matrix describes the mis-alignment between the SBES and the IMU. Therefore, the vector $C_{bS}^{bI} r_{bS}$ is the SBES return coordinated in the IMU frame.

Denoting by P_n the position delivered by the GNSS receiver (expressed in the navigation frame n), and X_n the sounding position we finally obtain the following spatial referencing equation:

$$X_n = P_n + C_{bI}^n \, (C_{bS}^{bI} \, r_{bS} + C_{bV}^{bI} a_{bI}) \qquad (1)$$

Spatial referencing error analysis purpose is to quantify the impact of measurement errors on the soundings X_n. Let us first differentiate between the positioning error and the ranging error. Indeed, any positioning error translates the sounding location. We can write $X_n = P_n + r_n$, where

$$r_n = C_{bI}^n \, (C_{bS}^{bI} \, r_{bS} + a_{bI})$$

In order to check the GNSS fix quality (i.e.; the precision of P_n) and in particular the effect of sea surface induced multi-path, the following procedure has been applied. The HydroBall® has been moored in the inter-tidal zone and GNSS data has been recorded during a tide cycle, as shown in Fig. 4. These static test concluded to the absence of variability of the GNSS position fix, as the horizontal and vertical errors were respectively 2.4 cm and 4 cm for 95 % of the observations, which is the

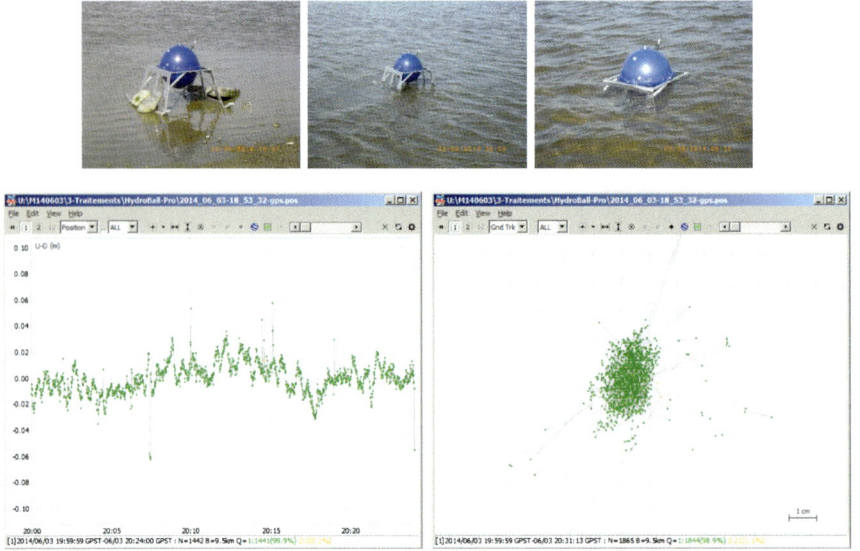

Fig. 4 On *top*, HydroBall® trials for assessment of the sea surface multipath refection. *Bottom* Vertical error through time and 2D horizontal error plots

same uncertainty level which was observed during static tests on geodetic control points.

The term r_n is formed by:

- The sounder return vector, expressed in the (n) frame: $r_n = C_{bI}^n \, (C_{bS}^{bI} \, r_{bS})$
- The lever-arm expressed in the (n) frame: $a_n = C_{bI}^n \, C_{bS}^{bI} \, a_{bI}$

We can write both r_n and a_n as a function of all sensors parameters:

$$r_n(\varphi, \theta, \psi, \delta\varphi, \delta\theta, \delta\psi, \rho) = C_{bI}^n(\varphi, \theta, \psi) \, C_{bS}^{bI}(\varphi_b, \theta_b, \psi_b) \, r_{bS}(\rho),$$

the term due to the ranging device and by

$$a_n(\varphi, \theta, \psi, a_x, a_y, a_z) = C_{bI}^n(\varphi, \theta, \psi) \, (a_x, a_y, a_z)^T,$$

the term due to lever arms. From (1), we have:

$$X_n(E, N, h; \chi) = P_n + a_n(\varphi, \theta, \psi, a_x, a_y, a_z) + r_n(\varphi, \theta, \psi, \varphi_b, \theta_b, \psi_b, \rho) \qquad (2)$$

Let us now denote by

$$\chi := (\varphi, \theta, \psi, \varphi_b, \theta_b, \psi_b, a_x, a_y, a_z, \rho)$$

the state vector of the HydroBall®.

The vector χ will be now supposed to lie within the neighborhood of any vector χ_0, and submitted to random uncertainty $\chi = \chi_0 + \delta\chi$, $\delta\chi$ being a random variable in R^8 with variance-covariance matrix $\Sigma_{\delta\chi}$.

We aim to propagate the variance/covariance matrix $\Sigma_{\delta\chi}$ through the geolocation equation. Unfortunately, the variance/covariance propagation law only applies to *linear transformations*,[1] we need to linearize equation (2). Linearization of (2) *around the measurement vector* χ_0 is nothing else than the Taylor expansion of X_n around χ_0:

$$R_n(\chi) - R_n(\chi_0) = \frac{\partial r_n}{\partial \chi}(\chi_0)\,(\chi - \chi_0)$$

Denoting by $\delta r_n = r_n(\chi) - r_n(\chi_0)$ and $\delta\chi = \chi - \chi_0$, we rewrite the previous equation by:

$$\delta R_n = \frac{\partial R_n}{\partial \chi}(\chi_0)\,\delta\chi \tag{3}$$

where $\frac{\partial f}{\partial \chi}(\chi_0)$ is the jacobian matrix[2] of X_n evaluated at point χ_0.

From (3) we can propagate the variance-covariance matrices of the measurement vector χ:

$$\Sigma_{\delta R_n} = \frac{\partial R_n}{\partial \chi}(\chi_0)\,\Sigma_{\delta\chi}\,\frac{\partial R_n}{\partial \chi}(\chi_0)^T \tag{4}$$

From this last equation, we can derive the variance of Easting, Northing and elevation of any sounding due to IMU and SBES measurements errors, lever-arms uncertainties. In addition to this, one should add the positioning error variance, leading finally to

$$\Sigma_{\delta X_n} = \Sigma_{\delta P_n} + \Sigma_{\delta R_n}$$

As an example, the a priori TPU has been computed in a particular configuration: $\phi, \theta, \psi = 20°$, $\phi_b = \theta_b = \psi_b = 0°$, $a_x = a_y = 0$, $a_z = 0.38$ m. The covariance matrix $\Sigma_{\delta\chi}$ is chosen directly according to sensors performances. Results are shown in Fig. 5.

---

[1] Let us recall that if $Y = AX$, X being a random vector with variance/covariance Σ_X, then $\Sigma_Y = A\,\Sigma_X\,A^T$.

[2] The jacobian matrix of a function $f : R^p \to R^n$ at point χ_0 is the linear operator represented by the matrix $[\frac{\partial f_i}{\partial \chi_j}(\chi_0)]_{ij}$.

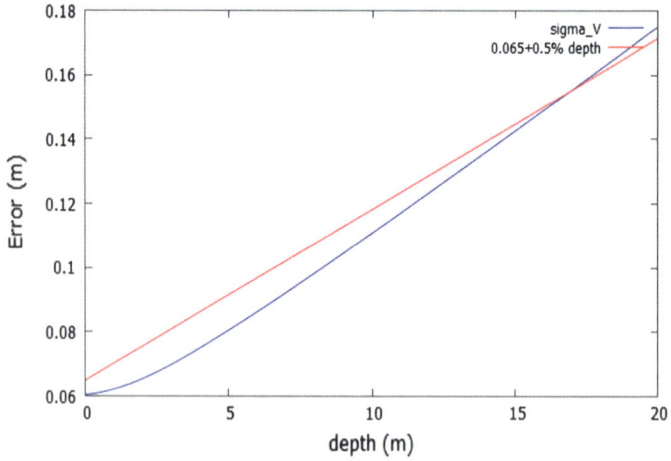

Fig. 5 Plot of the *horizontal* error and *vertical* error components versus a maximum admissible error bound defined for a particular application. From this plot we can see the maximum operational range of the system (about 17 m depth) in order to meet the uncertainty requirement

4 A Posteriori Total Propagated Uncertainty of the HydroBall® System

The analysis of the a posteriori TPU of the HydroBall® has been performed by using a reference surface constructed from a multi-beam survey conducted by the CIDCO in the Rimouski area, using a Reson 7125 MBES and a Pos-MV320/PPK hybrid inertial/GNSS positioning system. Figure 6 shows the reference surface and the surface constructed from HydroBall® data.

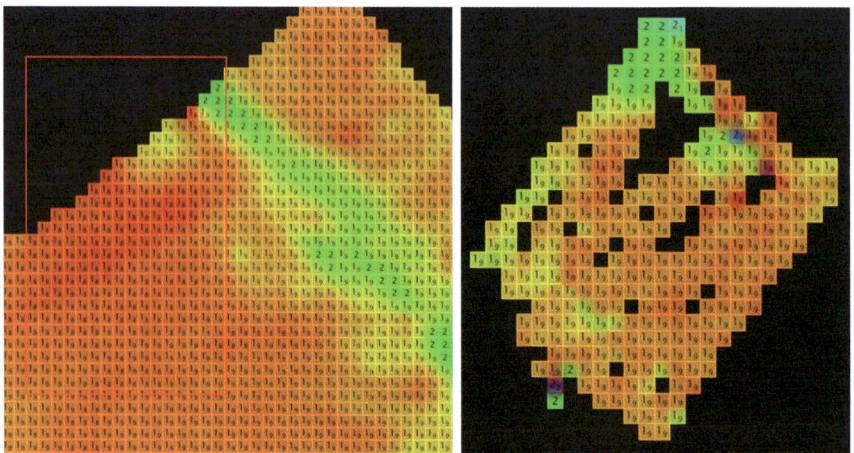

Fig. 6 MBES reference surface and Hydroball® data (on the *right*). The *red box* shows the location of the overlap between HydoBall and MBES datasets

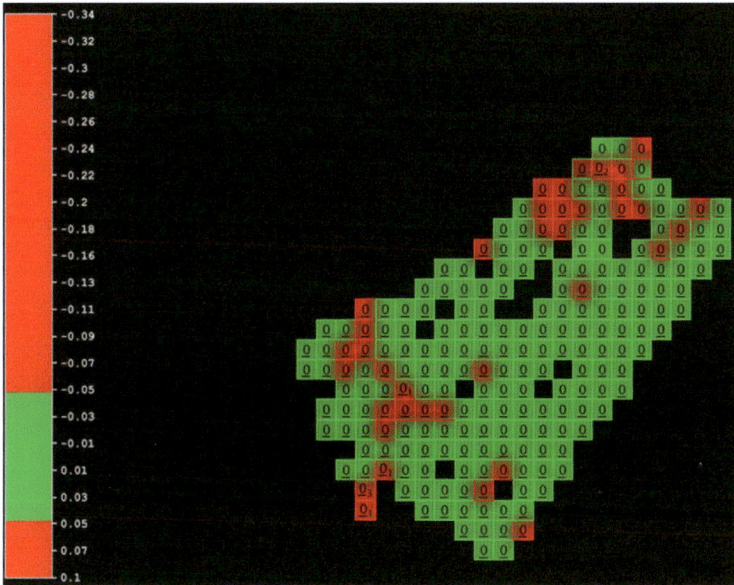

Fig. 7 Error surface between the HydroBall® dataset and the multibeam reference data set. Areas in *green* indicates an error lower than 5 cm. 95 % of the errors are less than 5 cm

The HydroBall® has been surveying the reference surface and an error analysis has been conducted for an area which average depth is about 5 m. We observed that 95 % of the error are less than 5 cm which is in accordance with the a priori error analysis, as shown in Fig. 7.

5   Conclusion and Future Work

This paper described an autonomous hydrographic survey buoy and shown the results that validate the data quality, according to international and industrial standards. First motivated by the survey of non accessible rivers, we shown that this system can be used in a flexible way for various applications. Its main advantage is that it does not require any survey ship installation and mobilization as it can be used on any opportunity boat or amphibious vehicle. As this system is compact, opened and offers open-source data processing tools, it is thus well adapted for hydrographers training. Indeed, all the principles of SBES data processing are implemented in a Python software, therefore enabling students to fully operate and understand SBES surveying activities.

Future work will focus on the real-time transmission of survey data by a wide range WiFi telemetry system and to the on-line quality control of survey data. The CIDCO developed quality control software tools devoted to single beam data analysis. They will be adapted to check in real-time the presence of systematic errors like erroneous sound speed profiles or positioning errors, in order to enable the remote user to monitor the data quality of the HydroBall® system.

References

1. Emery, L., Smith, R., McNeal, D., & Hughes, B. (2009). Drifting buoy for autonomous measurement of river environnment. In *OCEANS 2009, MTS/IEEE Biloxi—Marine Technology for Our Future: Global and Local Challenges*.
2. Swick, W., & MacMahan, J. (2009). The use of position-tracking drifters in riverine environments. In *OCEANS'09 MTS/IEEE Biloxi*.
3. MacMahan, J. (2010). *Drifter Trajectories in Riverine Environments*. Naval Postgraduate School, Monterey, CA, USA: Report Oceanography Department.
4. Emery, L., Smith, R., McQuary, R., Hughes, B., & Taylor, D. (2011, September 19–22). Autonomous river drifting buoys applications and improvements. In *OCEANS 2011*.
5. Albaladejo, C., Soto, F., Torres, R., Sanchez, P. & Lopez, J. (2012). A low-cost sensor buoy system for monitoring shallow marine environments. *Sensors, 12*(7), 9613–34.

Part II
Marine Robotics

Low-Power Low-Cost Acoustic Underwater Modem

Christian Renner, Alexander Gabrecht, Benjamin Meyer,
Christoph Osterloh and Erik Maehle

Abstract Recent advances in electronics and robotics enable automated and potentially unsupervised environmental underwater inshore monitoring with swarms of
small, low-power, autonomous underwater vehicles (AUV). To enable flexible and
self-organizing operation, underwater communication is required. However, existing solutions aim at long-distance communication, leading to high unit cost, high
power consumption, and large dimensions; hence rendering their application in robot
swarms practically and economically infeasible. In this paper, we present our ongoing development of a low-power low-cost underwater modem for acoustic communication. It features a low unit cost, small form factor, and low power consumption. It
is flexible, robust, and achieves a suitably high data rate and low transmission delay
for swarm coordination tasks.

Keywords Acoustic modem · Underwater communication · Low power · AUV ·
Swarm · MONSUN

1 Motivation

In the recent past, considerable advances in electronics and robotics have brought
forward stationary sensor networks, mobile underwater robots, and hybrid solutions. These techniques enable automated and potentially unsupervised environmen-

C. Renner (✉)
Institute smartPORT, Hamburg University of Technology, Hamburg, Germany
e-mail: christian.renner@tuhh.de

C. Renner · A. Gabrecht · B. Meyer · C. Osterloh · E. Maehle
Institute of Computer Engineering, Universität zu Lübeck, Lübeck, Germany
e-mail: gabrecht@iti.uni-luebeck.de

B. Meyer
e-mail: meyer@iti.uni-luebeck.de

E. Maehle
e-mail: maehle@iti.uni-luebeck.de

© Springer International Publishing Switzerland 2016   59
B. Zerr et al. (eds.), *Quantitative Monitoring of the Underwater Environment*,
Ocean Engineering & Oceanography 6, DOI 10.1007/978-3-319-32107-3_6

tal underwater inshore monitoring. Among the practical applications are water quality monitoring, structural monitoring, and the study of marine life [5].

In this application field, swarms of autonomous underwater vehicles (AUV) offer a monitoring solution that is flexible, reusable, and self-organizing. In past years, relatively small and inexpensive AUVs have been developed—e.g., the MONSUN robot in [6] has a length of 60 cm, a diameter of 10 cm, and an approximate unit cost of € 2 000. Its typical mission time is 5 h with an energy budget of 70 Wh. Based on such a platform, underwater inshore robot swarms can be put into praxis. To achieve autonomous and self-organizing swarm behavior, swarm members must be able to communicate underwater. In the envisioned scenario—with communication ranges of several meters and likely diffuse sight conditions—acoustic communication appears to be most suitable, since alternatives such as radio and optical communication usually suffer from low communication ranges. Existing commercial solutions of acoustic modems, however, aim at long-distance communication and therefore suffer from large dimensions, a unit cost of several thousand Euros, and a power consumption of several Watts (e.g., refer to [4]), hence rendering their application in robot swarms practically and economically infeasible.

To fill this void, we started the development of an acoustic underwater modem that meets the design criteria mandated by robot swarms consisting of tens of low-cost, low-power AUVs such as MONSUN. In particular, we aim at

- **low unit cost** to limit the impact on robot unit cost and hence enable economically attractive robot swarms,
- **small form factor** to allow for application in small AUVs such as MONSUN and comparable vehicles,
- **low power consumption** to impact the mission time as little as possible,
- **flexibility** to permit application-specific modifications,
- **high data rate and low transmission delay** to achieve fast response times for swarm coordination tasks,
- **robustness** against interference and packet loss.

In this paper, we present the architecture and the current status of our technical realization of the modem, discuss its integration into the MONSUN AUV, and point out the performance and characteristics of our latest prototype. We conclude the paper with a roadmap of our next steps.

2 System Design

To meet the requirements from Sect. 1, we carefully analyzed various design choices. With a particular focus on flexibility, we opted for a hybrid hardware/software solution. Here, the hardware is responsible for filtering and amplifying the analog acoustic signal, while the software controls the actual de-/modulation. It is hence possible to swap modulation and coding schemes to comply with the requirements of the application, e.g., regarding fault tolerance or bandwidth.

We chose incoherent binary frequency shift keying (BFSK), since it is well suited for communication between moving devices and keeps the hardware layout simple and cheap—e.g., it does not require extra hardware such as a phase-locked loop (PLL). To account for frequency shifts due to AUV movement, we selected a frequency-deviation of 200 Hz yielding a symbol duration of 2.5 ms. To elevate the data rate, we employ parallel transmission in up to five frequency bands with a carrier spacing of 600 Hz. To decrease the risk of inter-symbol interference caused by multi-path propagation, we apply frequency hopping spread spectrum (FHSS) with a hopping sequence of length five; hence shielding from echoes of up to 10 ms. Frequencies are orthogonal to prevent intra-symbol interference. The set of frequency bands and carriers is predefined but can be configured at runtime with configuration packets. Synchronization is achieved with a 16-symbol preamble in a separate frequency band. The length has been chosen to keep the overhead of transmission time low while ensuring reliable preamble detection. Channel activity is tracked in form of a received signal strength indication (RSSI) to provide means for multiple access schemes.

Choosing the frequency band for acoustic communication requires a trade off [8]: On the one hand, the acoustic channel is impacted by several noise sources, such as ships and animals, where noise frequencies are usually in sub or low kHz regions. Another important noise source are the thrusters of the AUV. On the other hand, the low-pass characteristic of the medium water limits the maximum frequency to roughly 100 kHz. We hence chose a frequency band from 14 to 30 kHz, resulting in a sufficient noise cancellation with a 16th order Sallen-Key band pass, organized as separate high-pass, low-pass, and amplifier elements for flexibility reasons.

We use several techniques to cope with channel interference. First, we employ single-error correcting and double-error detecting extended (7, 3) and (15, 4) Hamming codes. Second, interleaving is used to provide resilience against burst errors. Third, a 16-bit cyclic redundancy check (CRC-16) ensures data integrity.

3 Implementation and Integration

To perform all necessary de-/modulation steps with a sampling rate of 100 kHz (to comply with the Nyquist-Shannon sampling theorem while giving some potential back-up bandwidth) requires a relatively powerful microcontroller. To meet this end, we selected an Atmel AVR32UC3 [3] with a clock frequency of 66 MHz to enable real-time signal decoding of five bits in parallel. We built a first modem prototype based on an evaluation kit from Alvidi [1].

We recently designed a modem board tailored to the form factor of the MONSUN AUV. The board is shown in Fig. 1. It contains the microcontroller, power supply, status LEDs, and connectors for the filter chain and pre-amplifier (for receiving) and the power amplifier (for sending). Here, we opted for a flexible (but bulky) design to allow experimentation with different filter designs and topologies. We are currently analyzing and optimizing the filters and amplifiers.

Fig. 1 Components of the acoustic modem prototype (from *left* to *right*): corpus and mainboard expansion adapter for the MONSUN AUV, modem board, audio amplifiers (*upper row*) and filters (*lower row*)

Fig. 2 CAD drawing of the MONSUN AUV equipped with two hydrophones for acoustic communication

Hydrophones from Aquarian Audio [2] (models H1c and H2c) serve as transducers, since they provide a relatively constant transfer function in the used frequency range [7]. They are mounted on the left and right side of the MONSUN AUV via a corpus expansion adapter that also contains the modem board. Figure 2 shows a CAD drawing of the MONSUN AUV with this expansion adapter and the hydrophones installed.

The modem communicates with the host (i.e., the AUV) via serial line with a packet-based protocol. Packets received by the modem are transparently forwarded with the exception of configuration packets, which are identified by their packet type as part of the packet header. Among the configuration options are commands (packets) for changing the number of parallel bit transmission (by en-/disabling frequency bands) and adjusting the receive signal threshold for preamble detection.

4 Characteristics and Preliminary Evaluation

Our acoustic modem prototype has a total unit cost of approx. € 250 w.r.t. the sum of its component prices, of which € 180 are due to two hydrophones (we intend to cut costs by using a single hydrophone in the next revision). The electronic circuit has a near bank-card-sized circuit layout of 70 mm × 68 mm. The depth of the modem currently sums up to 40 mm with all filters and the audio amplifier installed. We intend to integrate the filters into the main board in a later revision, hence reducing height to approximately 5 mm. Flexibility is achieved through a modular filter design and software-based de-/modulation. It is hence possible to alter the filter chain and algorithms prior to deployment without modifying the modem board. The frequency setup can be changed during runtime via configuration packets.

Details about the evaluation of our first prototype can be found in [7]. It reports a power consumption of 530 mW for receiving and 770 mW for sending. Compared to the typical consumption of a MONSUN AUV, this corresponds to an overall power consumption of less than 6 %. The modem achieves a data rate of 2 kbit/s gross and ca. 1 kbit/s net for packets of at least 20 bytes (considering overhead due to encoding and synchronization). We carried out successful transmission tests for distances of up to 9 m—an appropriate value for the envisioned swarm scenario—in a pool environment with the first prototype. We will repeat those experiments with the new prototype shortly and expect similar results.

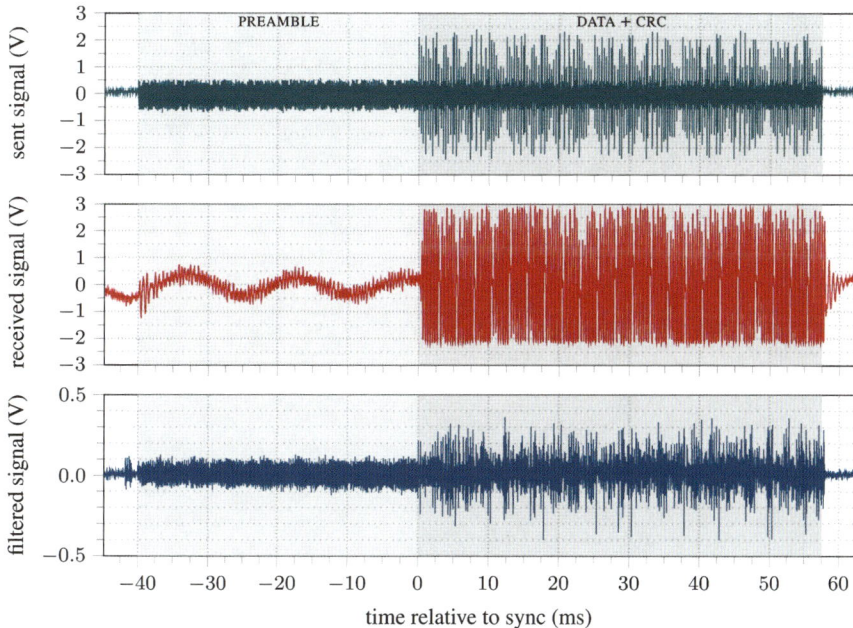

Fig. 3 Signal traces of a successful communication of two modems. Audio signal produced by the sending modem (*top row*) with the signal received by a second modem: unfiltered (*middle row*) and filtered. Traces were recorded with a Tektronix oscilloscope

In a first test series with our revised modem prototype, we verified successful packet transmission and reception in a small bucket (ca. 30 cm diameter) filled with water. We installed two hydrophones in the bucket, each connected to one of our prototypes. We used the only available new prototype as receiver and one of the older prototypes as sender. Please note that the sending circuitry of the new and old prototypes do not differ. Figure 3 shows an example comparison of the transmitted audio signal resulting from a short six-byte packet with the received and the amplified but unfiltered signal. We are currently evaluating motor noise of the MONSUN AUV and packet transmission/reception in a larger water tank.

5 Conclusion and Future Work

We presented our research towards a low-power low-cost acoustic underwater modem for use in small-sized AUVs. Our modem has been designed with a particular eye on enabling cheap but reliable underwater robot swarms. Due to its modular design and the use of software solutions rather than hardware implementations, where possible, the modem can be easily modified for different scenarios and application requirements. This realization is also practical for fine-grained performance evaluations and possible modifications of its components.

During the design and evaluation of the current modem revision, we have identified several challenges that we plan to tackle as future work. Up to now, we have only been able to run communication tests in a controlled indoor, pool environment. We plan to run thorough evaluations in an inshore lake and a small harbor site. Here, we will particularly concentrate on directionality, noise cancellation, typical bit error ratios, and the transfer characteristics of the individual carriers. The results will be used to fine-tune the input signal amplification and filter chain. We also aim at using a single hydrophone for both transmission and receiving to further reduce costs and power consumption. To extend the communication range and to achieve a high packet success rate, we plan to explore methods to adapt the amplification of the acoustic signal automatically.

Acknowledgments Our research was partially funded by the Business Development and Technology Transfer Corporation of Schleswig-Holstein (WTSH) as part of the initiative *Sea our Future*, project number 122-13-005.

References

1. ALVIDI: AVR32-Development Module (Model: AL-UC3AEB). Retrieved October 09, 2014, from http://alvidi.de/avr32_module.html.
2. Aquarian Audio Products: Aquarian Audio hydrophones. Retrieved October 06, 2014, from http://www.aquarianaudio.com/hydrophones.html.

3. Atmel Corp.: Atmel AT32UC3A1512. Retrieved October 06, 2014, from http://www.atmel. com/devices/AT32UC3A1512.aspx.
4. Evologics GmbH: Evologics Underwater Acoustic Modems. Retrieved April 25, 2014 from http://www.evologics.de/en/products/acoustics/.
5. Heidemann, J., Stojanovic, M., & Zorzi, M. (2012, January). Underwater sensor networks: applications, advances, and challenges. *Philosophical Transactions of the Royal Society-A, 370*(1958), 158–175.
6. Meyer, B., Ehlers, K., Isokeit, C., & Maehle, E. (2014). The development of the modular hard- and software architecture of the autonomous underwater vehicle MONSUN. In *Proceedings of the 45th International Symposium on Robotics (ISR 2014) and 8th German Conference on Robotics (ROBOTIK 2014)*, June 2014.
7. Osterloh, C. (2013, February). Ein akustisches Modem für die Kommunikation in Schwärmen autonomer Unterwasserroboter. Ph.D. thesis, Institute of Computer Engineering, Universität zu Lübeck, Lübeck.
8. Stojanovic, M., Preisig, J. (2009, January). Underwater acoustic communication channels: prop- agation models and statistical characterization. *IEEE Communications Magazine*, 84–89.

IMOCA: A Model-Based Code Generator for the Development of Multi-platform Marine Embedded Systems

Goulven Guillou and Jean-Philippe Babau

Abstract Process control systems embedded in disturbed environments are usually developed case by case for specific deployment platforms and their behaviours closely depend on the characteristics of the environment. The obtained code is not portable and not reconfigurable. In order to help the software development of such applications, IMOCA offers architectural modelisation tools. The associated code generator allows to product adaptive and reconfigurable code for a simulator as well as embedded code for various platforms. This approach has been tested on NXT bricks, Arduino boards and Armadeus boards.

Keywords Software architecture · Control · Code generation

1 Introduction

Embedded systems in an unpredictable and disturbed environment, like underwater control systems, have to take into account various situations by considering different strategies. Their development requires large parameters configuration in order to ensure safety and efficiency of the controlled system.

The configurable parameters are used to characterize the context (environment interpretation and its evolution), the execution platform and the control part of the system. The tuning of the parameters is based on simulations for cost reasons and on real testing for safety reason. At the end, the system has to be equipped with adaptation and/or learning abilities to adjust some parameters online. Therefore, there is a strong need for offline and online tuning tools. Unfortunately in industry, the development of the software for such systems is mainly focused on the code efficiency. The produced embedded code is dedicated to a given platform for a specific appli-

G. Guillou (✉) · J.-P. Babau
Lab-STICC/UMR 6285, UBO, UEB, 20 Avenue Le Gorgeu, 29200 Brest, France
e-mail: goulven.guillou@univ-brest.fr

J.-P. Babau
e-mail: jean-philippe.babau@univ-brest.fr

© Springer International Publishing Switzerland 2016   67
B. Zerr et al. (eds.), *Quantitative Monitoring of the Underwater Environment*,
Ocean Engineering & Oceanography 6, DOI 10.1007/978-3-319-32107-3_7

cation in a specific context. The obtained code is difficult to maintain and to adapt for new applications and new contexts. Portability and reusability are limited.

To tackle these limitations, we propose in [1] a model-based software architecture integrating high adaptive capabilities, the IMOCA approach. Once the architecture model is established, a key point is then the code generation. In this paper, we present the implementation of the model-based code generator from the architecture model IMOCA. Even if the deployed code remains specific, we propose generic models of code generation to allow:

- the control of quality and efficiency of the generated code: the code is optimized for static parts and modular for adaptive part;
- the integration, and so the reuse, of existing specific legacy code is facilitated: domain functions such as control law, communication protocol, acquisition policy can be easily integrated;
- the code generation for different target platforms: the architecture model and most of the code generator is independent of a given platform, specific aspects are encapsulated in a platform abstraction view;
- the model and the code generator integrate testing, adaptation and tuning facilities: design facilitates declaration of adaptive parameters and functions, parameters can be tested with a generated simulator, code integrates reconfiguration capabilities for adaptive parts.

The generated simulator is written in Java, while the generated embedded code is written in C language (or family of C language). To generate optimized embedded code, the code generator is based on the principles presented in [2]. The models and tools have been tested on NXT bricks, Arduino boards and Armadeus boards for simple control applications, sand yacht control and autonomous sailing boat VAIMOS control.

After a presentation of the IMOCA architecture model, we present the structure and the underlying principles of the code generator while using it with two applications on two different execution platforms.

2 Related Works

Component-based approaches [3] for the conception of the embedded systems are relatively classical and allow to deal with the software complexity [4] and the well-known *separation of concerns* in Model Driven Architecture (MDA) terminology [5]. In particular, software evolution ability in order to take into account either the platform maintenance or system behavior adaptation or parameters tuning [6] can be viewed as software flexibility [7]. Generally, efforts to increase the flexibility lead to conflicts with respect to the material resources of the execution platform [8]. In [9] binding each component to an adaptation policy allows software evolution, but this solution assumes that the evolutions are predictable. [2] tries to conciliate the software evolution at *run-time* with the hard resource constraints of the embedded platforms. Relying on this work, we focus on the conception of *domain-specific* models of components (we focus on a specific domain without modeling specific features of the components), then we generate optimized code which embeds the necessary elements for the reconfiguration [2].

3 The Architecture Model IMOCA

IMOCA for archItecture for MOde Control Adaptation is an architecture model dedicated to the development of process control systems embedded in a disturbed environment. This architecture is composed of three layers called `Target`, `Interpretation` and `Control` (see Fig. 1). `Target` with its `Actuators` and `Sensors` is a platform-specific model of I/O. `Control` uses `Data` (an ideal view of the environment) to compute ideal `Commands` to act on the environment. The `Interpretation` layer realizes the adaptation between the `Target` layer and the `Control` layer by linking `Sensors` and `Actuators` on the one hand, to `Data` and `Commands` on the other hand. In this way `Control` and `Target` are independent like in SAIA [10] and this allows the development independently of specific sensors and actuators technologies. This independence is important in the context o f embedded systems because material platforms may be various and may evolved (change or add a sensor for example).

The `Controller` is composed of three sub-controllers. The `Reactive Controller` applies a control law to compute a `Command` from `Data`. The `ExpertController` is in charge of defining the current control law. It is based on a finite state automaton that manages running modes. Each state is associated with a `Mode` which is itself associated with a control law. A state change is linked to a change of state of the environment (a function of `Data` which returns a boolean). Finally, an `AdaptativeController` adjusts different parameters of the control law with respect to a look-up table in which appear all the possible `Configuration`. Based on this three collaborating controllers, the `Controller` allows to answer the three following requirements: controlling the process with

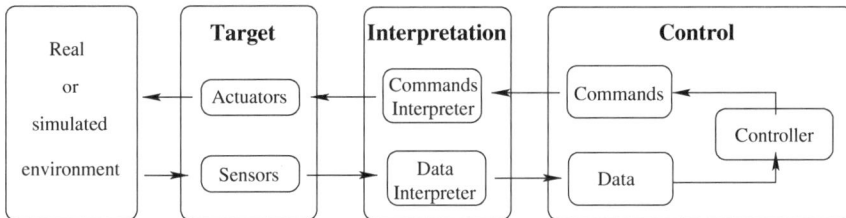

Frames represent components, arrows describe data flow

Fig. 1 Principles of IMOCA approach

the `ReactiveController` by applying an adapted control law thanks to the `ExpertController`, and finally, adjusting the control laws with respect to the context in order to keep a high quality of control with the `Adaptative Controller`.

4 Code Generation

The model of the software architecture design is an instance of the meta-model IMOCA, expressed using ecore. The code generator leans on Acceleo technology which allows to define code generation principles through an Ecore meta-model.

4.1 Parameters Configuration

A software application based on the architecture model IMOCA integrates variability to adapt the system to a specific context or a specific platform. Variability is first implemented through parameters to characterize, for example, a threshold in a filter adapter, or a coefficient for a control law. Second, the `ExpertController` automaton defines a dynamic behavior through running modes. To tune parameters and modes, we propose to generate simulation tools to evaluate the impact of different parameter values and actions on the system.

In this paper, we target the generation of an adaptive and reconfigurable embedded code. "Adaptive code" means the application is able to take into account the evolution of its environment through the `adaptativeController`. For each environment context, specific values of parameters are defined at design time using the simulation. Then, the `adaptativeController` adapts online the values depending on the context. "Reconfigurable code" means that the embedded values of the parameters can be modified online without the need to recompile the code. The reconfiguration capability concerns parameters only, the architecture of the application cannot be modified online.

In this version, a Java simulator is generated. The generation of this simulator is based on high-level data (sensors and actuators are not considered here) and includes all the controllers. The designer can test different control laws and parameter setting through a dedicated generated User Interface. Each `Data` can be controlled and each new `Command` is printed on a control screen. To view the real effect of the controller, it is necessary to implement an environment simulator for the system itself (as in [10]). The latter must be connected to the commands sent by the generated simulator.

4.2 Taking into Account the Behavior

The architecture model IMOCA is a declarative model. It allows to focus on specific features of the target application (activation periods, number of operating modes,

...). However, IMOCA is not a programming language, the expected behavior is not described explicitly. The expected behavior is implemented by the code generator, following the IMOCA semantic.

For this purpose we distinguish two parts. The former corresponds to the operational semantic of IMOCA which has been given in the previous section. The code generator is directly in charge of this part. The behavior is expressed in a simple C code (no pointer) to be reusable in different languages. The latter is specific to application-dependent part and is encapsulated in domain-specific libraries. The code generator is in charge to produce glue code to link these two parts *via* a simple library integration.

According to these principles, the `ExpertController` and the `adaptive Controller` automata are generated together with a general controller and the code dedicated to `Data` and `Command`. For the `Sensors` and the `Actuators`, the `Interpreters` and the `ReactiveController`, only the declaration and glue code is generated. To complete the code, the user provides specific libraries to deal with sensors acquisition (filters, ...), actuators management, Interpretation and control laws.

As an example, the generated code presented Fig. 2 defines the expert controller of a mini sand yacht and is, in fact, the implementation of a finite state automaton. Each state change is bound to the evaluation of the data `theta` which represents the heeling of the sand-yacht in degrees. For information, here we have three states (states 1, 2 and 3) respectively corresponding to a normal state, a (excessive) heel to starboard and a heel to port.

4.3 Domain-Specific Code Integration

As previously said, generated code must be completed by adding domain-specific code. Since many dedicated libraries exist, the idea is to generate code in a way that allows a smooth integration. As in a component-based approach [5], IMOCA produces a set of signatures of functions for all relevant components (sensors, actuators, filters and reactive controller). Thus, the designer has to provide the set of corresponding implementations (user code) while respecting the static typing. In addition, we use the properties of Acceleo to add, in the generated code, portions of customizable code. For this purpose, a special place is pointed by a commentary and is available for the user (see below). By default, a code is provided which can be modified and completed. It should be noted that subsequent generations of code take into account these changes (automatically performed by Acceleo). This default code calls a user function for each domain-specific component, user function to be implemented.

```
bool UpdateBoolSensorTouch(){
  bool aBoolSensorTouch;
  // Start of user code for ReadBoolSensorTouch definition
```

```
#ifndef EXPERT
#define EXPERT

/* file expert controller */

int state = 1;

int getState(int theta) {

switch(state)
{
case 1 :
if (theta > 20) { state = 2; }
if (theta < -20 ) { state = 3; }
break;
case 2 :
if (theta < -20 ) { state = 3; }
if (theta >= -20.0 && theta <=20) { state = 1; }
break;
case 3 :
if (theta > 20) { state = 2; }
if (theta >= -20.0 && theta <=20) { state = 1; }
break;
}
return state;
}

#endif
```

Fig. 2 Code of the expert controller

```
aBoolSensorTouch=UserReadBoolSensorTouch(touchPort);
// End of user code
boolSensorTouch = aBoolSensorTouch;
return aBoolSensorTouch;
}
```

If the user keeps the call of the specific function, he must provide a function that respects the signature required by the code generator to the specific library:

```
bool UserReadBoolSensorTouch(int port) {
    return ((Sensor(port)==0)) ;
}
```

4.4 Taking into Account Platforms

In order to address various execution platforms, we need to generate specific code for each of them. However, a significant part of the code generator must remain generic and independent of the target language. For example, the behavior of the `ExpertController` is independent of the target language.

To address this requirement, the code generator is based on three software layers. The first one is concerned with generating the specific behavior related to IMOCA. It can generate imperative code or object-oriented code. We add a parameter to each generative function to define the expected code (imperative or object-oriented). The second layer is based on the first one and is specific to the target language (Java or C for example). It is in charge of generating the files respecting the specific language features for the declaration of files, classes or functions. Currently, the specificities of the executive are included in the second layer *via* the definition and the call of tasks. The last layer is in charge of general services.

For the first version of the code generator, the target language is based on the basic constructs of C language (assignment, control structures). Pointers are not used. Thus, we can rely on this layer to generate C, C++, Java code or any C-like language.

We have used the code generator for two applications deployed on different platforms. The former is a NXT Lego brick equipped with a motor and a touch sensor. We just control the speed of rotation of the motor by using the touch sensor. The latter is a mini sand yacht with an Arduino board Mega and an inertial motion unit (IMU). We have to try to keep the course and, in the same time, to avoid the capsize of the vehicle due to the wind action.

At the level of second layer, the first part of the code presented Fig. 3 allows to generate the file `Input.nxc` whereas Fig. 4 presents the same thing for the simulator (some lines of commentary have been removed).

We generate NXC source code for the NXT platform. NXC means Not eXactly C, a C-like language with some specific features. Generated files have an extension of `nxc` (here the files are `Input.nxc` and `Output.nxc`) and we retrieve the preprocessor invocations like in C. We `include` only implementation files, because NXC does not h integrate interface files. Each data needs a declaration, the definition of their attributes (value, frequency, format ...) and of their access methods.

The equivalent Acceleo code to generate the simulator (see Fig. 4) produces Java code. A `Data` is viewed as a high-level data, that is to say here as a class. However calls like `[generateWriteDefinition(data,0)/]` remains identical to those used to generate code for the NXT brick.

In the first layer, Acceleo modules are parameterized by the type of the used language (`0` for object-oriented programming, and `1` for imperative programming). Figure 5 shows this case.

This leads for the case of the NXT brick (the commentaries have been removed) to:

```
void InitControls() {
InitGo();
```

```
Init_Stop();
}
```

whereas the generated Java code is:

```
public void InitControls() {
myReactiveController.InitGo();
myReactiveController.Init_Stop();
}
```

```
[template public generateInputOutputNXC(aSystem : System)]

[file ('Input.nxc', false, 'UTF-8')]

#ifndef INPUT
#define INPUT

/* file for inputs */

[for(data : Data | aSystem.data->select(oclIsTypeOf(EnumeratedInput)
                                    or oclIsTypeOf(ContinuousInput)))]
[generateDataAttributeDeclaration(data,0)/]
[/for]

[for(data : Data | aSystem.data->select(oclIsTypeOf(EnumeratedInput)
                                    or oclIsTypeOf(ContinuousInput)))]
[generateWriteDefinition(data,0)/]
[generateReadDefinition(data,0)/]
[/for]

#endif

[/file]

[file ('Output.nxc', false, 'UTF-8')]

#ifndef OUTPUT
#define OUTPUT

/* file for outputs */

#include "OutputToActuator.nxc"
```

Fig. 3 Fragment of the generateInputOutputNXC.mtl Acceleo module

```
[template public generateInputOutputJava(aSystem : System)]
[for(data : Data | aSystem.data->select(oclIsTypeOf(EnumeratedInput)
                                 or oclIsTypeOf(ContinuousInput)))]
[file (name.toUpperFirst().concat('.java'), false, 'UTF-8')]

/**
 *  Class for [name/] data
 */

package gener[aSystem.name/];

public class [name.toUpperFirst()/] {

protected [generateDataAttributeDeclaration(data,0)/]

public [generateWriteDefinition(data,0)/]

public [generateReadDefinition(data,0)/]
}
[/file]
[/for]
```

Fig. 4 Fragment of the generateInputOutputJava.mtl module

```
[template public generateInitControlsDefinition
                               (aSystem : System, lang : Integer)]
void InitControls() {
// [protected ('for Init Adapters Definition')]
// user code
// [/protected]
[for(control : Control | aSystem.modes.command)]
[if (lang=0)]myReactiveController.[/if]
                               Init[control.name/]();
[/for]
}
[/template]
```

Fig. 5 Acceleo template for generating object-oriented and imperative code

4.5 Structure of the Code Generator

The code generator is modular with a set of specific modules for each component
of IMOCA. For the first and second layers, a code generation module is proposed
for each modeled entity (Sensors, ExpertController, ...). Utilities modules
are used to complete the code generator in order to factorize and simplify common
pieces of code.

```
[template public generateWriteDeclaration(data : Data, lang : Integer)]
void Write[name.toUpperFirst()/]
    ([if(data.oclIsKindOf(EnumeratedData))]int
    [else]
        [syntaxType(data.oclAsType(ContinuousData).type,lang)/]
    [/if] value) ;
[/template]
```

Fig. 6 Fragment of the generateDataUtils.mtl Acceleo module

For the first layer, simple components (`ReactiveController`, `Sensor`, `Actuator` and `Interpreter`) are built according to the same principles. A class (or a structure in the case of imperative code) is generated which manages objects (or structure) that contain typed data, an initialization function, a configuration function, a getter, a setter and specific executive functions (a "run" for each filter and for each control law of reactive controller). The other control components are generated through a specific function which implements automata. Other components implement a service layer (to respect the `Facade Design Pattern`, a main program (the *main*) and communication tools (to implement communication with a remote client in charge of reconfiguration).

Figure 6 shows a piece of Acceleo code which stands in the first layer of the generator and concerns the `Data`.

To assist testing, the generated code is also modular. For object-oriented language, a class is generated for each element (each sensor, each data, each controller, …). For an imperative language, a file is generated for each element. Thus, the structure helps on unit testing activity.

4.6 Reconfiguration

The code generator generates reconfigurable code in the sense that parameter values must be changed online. To prepare the code generation, each parameter can be recorded as reconfigurable (the value of the property `IsControllable` which is `false` by default, is set to `true`). If at least one reconfigurable parameter exists, a client interface (currently in Java) is generated to allow the tuning of the values of reconfigurable parameters. In the embedded code, a server task retrieves the changes and modifies online the corresponding parameters (this modification is done *via* a call to the corresponding setter). This tool is especially useful during prototyping phase. The behavior of the system can be tested without having to stop and recompile the whole application [2].

5 Conclusion and Future Works

This paper presents an approach to generate adaptive and reconfigurable embedded code based on the architecture model IMOCA. The code is designed for process control systems in disturbed environments and can be generated for different platforms. A Java simulator is generated to assist the user in the tuning of control laws. The case of an NXT robot and a mini sand-yacht equipped with an Arduino board have been used for experiments.

We are working on addressing other platforms and on optimizing the code to take into account limited platforms. We also seek to control the consumption of autonomous systems, like drones, by using control policies based on energy criteria. Finally, we seek to enrich the integration of existing specific libraries to deal with other application areas.

References

1. Guillou, G., & Babau, J. P. (2013). IMOCA : une architecture à base de modes de fonctionnement pour une application de contrôle dans un environnement incertain. In 7ème Conférence francophone sur les architectures logicielles. Toulouse France.
2. Navas, J., Babau, J.-P., & Pulou, J. (2013). Reconciling run-time evolution and resource-constrained embedded systems through a component-based development framework. *Science of Computer Programming, 8*, 1073–1098.
3. Szyperski, C., Gruntz, D., & Murer, S. (2002). *Component software: Beyond object-oriented programming*. New York: ACM Press and Addison-Wesley.
4. Crnkovic, I. (2005). Component-based software engineering for embedded systems. In *Proceedings of the 27th International Conference on Software Engineering, ICSE 05* (pp. 712–713). New York, NY, USA: ACM.
5. Anne, M., He, R., Jarboui, T., Lacoste, M., Lobry, O., Lorant, G., et al. (2009). Think: View-based support of non-functional properties in embedded systems. In *ICESS 09: Proceedings of the 2009 International Conference on Embedded Software and Systems* (pp. 147–156). Washington, DC, USA: IEEE Computer Society.
6. Chapin, N., Hale, J. E., Khan, K. M., Ramil, J. F., & Tan, W.-G. (2001). Types of software evolution and software maintenance. *Journal of Software Maintenance and Evolution: Research and Practice, 13*(1), 3–30.
7. Mathieu, J., Jouvray, C., Kordon, F., Kung, A., Lalande, J., Loiret, F., et al. (2012). Flex-eWare: A flexible MDE-based solution for designing and implementing embedded distributed systems. *Software: Practice and Experience, 42*(12), 1467–1494.
8. Navas, J., Babau, J.-P., Lobry, O. (2009). Minimal yet effective reconfiguration infrastructures in component-based embedded systems. In *Proceedings of the ESEC/FSE Workshop on Software Integration and Evolution @ Runtime (SINTER09)*.
9. Borde, E., Haik, G., & Pautet, L. (2009). Mode-based reconfiguration of critical software component architectures. In *Design, Automation Test in Europe Conference Exhibition, 2009. DATE 09* (pp. 1160–1165).
10. DeAntoni, J., & Babau, J.-P. (2005). A MDA-based approach for real time embedded systems simulation. In *Proceedings of the 9th IEEE International Symposium on Distributed Simulation and Real-Time Applications* (pp. 257–264). Montreal: IEEE Computer Society

Visual Servoing for Motion Control of Coralbot Autonomous Underwater Vehicle

Eduardo Tusa, Neil M. Robertson and David M. Lane

Abstract This work focuses on developing two visual servoing algorithms that enable the Coralbot to stabilise itself relative to an area of reef. For this purpose, we present a fast coral reef detector based on supervised machine learning. We extract texture feature descriptors using a bank of Gabor Wavelet filters. We use a database of 621 images of coral reef located in Belize. The Decision Trees algorithm shows a fast execution time among the machine learning algorithms. We use the coral detections to estimate point features and moment features. We use these features through an Image-based approach and a Moment-based approach. We code the coral detector and the visual servoing algorithms in C++ for obtaining a fast response of the system. We test the system performance through an underwater simulator, UWSim, which is supported by the Robot Operating System, ROS. We obtain promising results using point features instead of moment features.

Keywords Coralbot · Coral reef · Machine learning · Gabor Wavelet filters · UWSim

1 Introduction

The conventional methods of coral reef restoration involve extreme conditions for volunteer SCUBA divers, who transplant loose fragments back onto the larger reef framework. *Lophelia pertusa* is one of the most important reef-building coral species

E. Tusa (✉)
Universidad Técnica de Machala, Machala, Ecuador
e-mail: etusa@utmachala.edu.ec

N.M. Robertson · D.M. Lane
Heriot-Watt University, Edinburgh, UK
e-mail: N.M.Robertson@hw.ac.uk

D.M. Lane
e-mail: D.M.Lane@hw.ac.uk

© Springer International Publishing Switzerland 2016
B. Zerr et al. (eds.), *Quantitative Monitoring of the Underwater Environment*,
Ocean Engineering & Oceanography 6, DOI 10.1007/978-3-319-32107-3_8

79

in the world. However, its location generates limitations to the human intervention due to the low temperatures and the deep sea, which is around 200 m [1].

The Coralbot project is a recently proposed idea to autonomously repair deep-sea coral reefs and thus support the oceans ecosystem which is vital for commercial fishing, tourism and other species. The idea consists of combining autonomous underwater robots with swarm intelligence, which mimics the behaviour of organisms (bees, termites and wasps) acting in group and performing complex tasks; just by following simple rules. Thus, this team of AUVs are deployed to recognise coral reef and execute restoration tasks.

We develop two visual servoing algorithms by testing through an underwater simulator. The algorithms use the features provided by a fast coral reef detector based on machine learning algorithms. These algorithms use Gabor Wavelet filters to extract texture feature descriptors. The coral detector is integrated to the Coralbot, whose prototype results of fusing Nessi VII from Ocean Systems Lab with a robot arm ARM5. Nessie VII is a torpedo shape vehicle that has 5 degrees of freedom (dof), 6 thrusters, which are controlled by sending velocity/displacement commands to a low level controller. ARM5 is a robot arm, in which we use 4 dof that correspond to revolute joints.

Next, we discuss the state of the art, in which we summarise the main theories and current researches that explain the detection of coral reef and the visual servoing approaches. Then, we illustrate the development of the algorithms used in this project. The next section explains the results of the algorithms of coral detection and visual servoing. Finally, we present the conclusions.

2 State of the Art

2.1 Coral Detector

2.1.1 Feature Descriptors

A considerable amount of literature has been published on feature extraction. Mainly, the image is transformed into a set of feature vectors, so that various desired regions or shapes are described quantitatively by their properties (colour intensity, texture information, spatial data, edge cues).

The type of features used in most of the research papers are based on colour [2, 3] and texture [3–5] information. Purser et al. [6] compute 15 differently oriented and spaced gratings in order to produce a set of 30 texture features, and to compare a computer vision system with the use of three manual methods: 15-point quadrat, 100-point quadrat and frame mapping.

Colour features are sensitive to the lack of illumination on the seabed. For this reason, we extract texture feature descriptors using the Gabor Wavelets filters used by Purser et al. [6] but, we implement seven scales and five orientations [7].

2.1.2   Discrimination Algorithms

The design of the algorithm oriented to the discrimination of classes: coral and non-coral, is addressed in several ways. Most of the previous papers, authors apply the idea of machine learning by using different techniques for classification [8]. We take a feature vector for every pixel of the image and we assign a class: coral and non-coral. The assignation of a feature vector is fitted according to a prediction model, that has been derived from training data.

Purser et al. [6] develop a coral detector using Neural Networks. The results are satisfactory, but the algorithm takes a significant time for coral detection. This amount of time for processing a frame is an obstacle for real-time applications.

For this reason, we compare the running time and accuracy of nine machine learning algorithms such as: Decision Trees (DTR) [9], Random Forest (RTR) [10], Extremely Randomised Trees (ERT) [11], Boosting (BOO) [12], Gradient Boosted Trees (GBT) [13], Normal Bayes Classifier (NBA) [8], Expectation Maximisation (EMA) [14], Neural Networks (MLP) [10], Support Vector Machines (SVM) [8].

2.2   Visual Servoing

Vision-based control uses computer vision data in order to incorporate corrective actions to the robot motion [15]. These actions are implemented by using a speed controller that pursues the system stability.

If we define the camera velocity $\xi(t)$ and $s(t)$ represents a set of image features, the relationship between $s(t)$ and $\xi(t)$ is given in (1)

$$s(t) = L(s, q)\xi(t) \tag{1}$$

where $L(s, q)$ is known as the interaction matrix or Image Jacobian matrix. The goal configuration of image features is denoted by s^d. Thus, the image error function is defined in (2)

$$e(t) = s(t) - s^d \tag{2}$$

A control law that estimates the camera velocity $\xi(t)$ is expressed as follows in (3)

$$\xi(t) = \lambda L^+ e(t) \tag{3}$$

where $L^+ = (L^T L)^{-1} L^T$ is the pseudoinverse of the interaction matrix, and λ represents a proportional gain. The visual servoing algorithm follows the scheme presented in Fig. 1a. From our vision sensor, we get an image $I(t)$ that is processed by our computer vision system, which extracts features $s(t)$ (points or moments). These are compared to our desired features s^d in order to generate an error function $e(t)$.

(a)

(b)

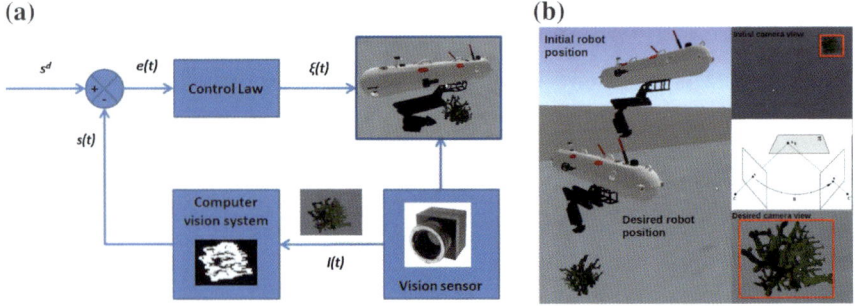

Fig. 1 **a** Block diagram of the visual servoing algorithm. **b** Task of station keeping

We use this error to calculate the velocity $\xi(t)$ that is executed by robot. Coralbot performs a task of station keeping relative to coral reef. Figure 1b shows the initial and desired robot positions.

2.2.1 Image-Based Approach

The image data is used directly to control the robot motion. It is very common to use detected points on an object as feature points. The interaction matrix for a set of 4 points is given in (4)

$$
L = \begin{bmatrix}
-\frac{1}{Z_1} & 0 & \frac{x_1}{Z_1} & x_1 y_1 & -(1+x_1^2) & y_1 \\
0 & -\frac{1}{Z_1} & \frac{y_1}{Z_1} & (1+y_1^2) & -x_1 y_1 & -x_1 \\
\vdots & \vdots & \vdots & \vdots & \vdots & \vdots \\
-\frac{1}{Z_4} & 0 & \frac{x_4}{Z_4} & x_4 y_4 & -(1+x_4^2) & y_4 \\
0 & -\frac{1}{Z_4} & \frac{y_4}{Z_4} & (1+y_4^2) & -x_4 y_4 & -x_4
\end{bmatrix}
\tag{4}
$$

where the features $s_i = (x_i, y_i)$ are the point coordinates expressed in the image coordinate frame, and Z_i is the depth with respect to target from which we extract features.

2.2.2 Moment-Based Approach

For a discrete set of n image points, the moments are defined by

$$
m_{ij} = \sum_{k=1}^{n} x_k^i y_k^j
\tag{5}
$$

and the centered moments are given by

$$
\mu_{ij} = \sum_{k=1}^{n} (x_k - x_g)^i (y_k - y_g)^j
\tag{6}
$$

where the coordinates of the center of gravity (x_g, y_g) are described by

$$x_g = \frac{m_{10}}{n}, \; y_g = \frac{m_{01}}{n}, \; m_{00} = n$$

The moments centered are characterised to be invariant to 2D translational motion. In [16], the author presents several combinations of moments $(c_1, \ldots, c_{10})$, that are invariant to 2D translation, 2D rotation and to scale. These moments are selected as visual features to control rotational velocities ω_x and ω_y.

In [17], the coordinates x_g, y_g and the area $a = m_{00}$ of the object in the image are used to control the translation dof. Tahri et al. [16] introduces a normalisation that generates a complete partition of these three selected features. Thus, they define three identities in (7)

$$a_n = Z^* \sqrt{\frac{a^*}{a}}, \; x_n = a_n x_g, \; y_n = a_n y_g \tag{7}$$

where a^* is the desired area of the object in the image, Z^* is the desired depth between the camera and the object. The object orientation α [17] and two invariants moments c_i and c_j are selected to control the rotational dof. Thus, the interaction matrices related to these normalised features can be obtained as follows in (8)

$$L = \begin{bmatrix} -1 & 0 & 0 & a_n\epsilon_{11} & -a_n(1+\epsilon_{12}) & y_n \\ 0 & -1 & 0 & a_n(a+\epsilon_{21}) & -a_n\epsilon_{22} & -x_n \\ 0 & 0 & -1 & -a_n\epsilon_{31} & a_n\epsilon_{32} & 0 \\ 0 & 0 & 0 & c_{i_{wx}} & c_{i_{wy}} & 0 \\ 0 & 0 & 0 & c_{j_{wx}} & c_{j_{wy}} & 0 \\ 0 & 0 & 0 & \alpha_{wx} & \alpha_{wy} & -1 \end{bmatrix} \tag{8}$$

where

$$\epsilon_{11} = \frac{\mu_{11}}{m_{00}} + x_g(y_g - \epsilon_{31}), \epsilon_{12} = \frac{\mu_{20}}{m_{00}} + x_g(y_g - \epsilon_{32})$$

$$\epsilon_{21} = \frac{\mu_{02}}{m_{00}} + y_g(y_g - \epsilon_{31}), \epsilon_{22} = \frac{\mu_{11}}{m_{00}} + y_g(x_g - \epsilon_{32})$$

$$\epsilon_{31} = y_g + \frac{y_g\mu_{02} + x_g\mu_{11} + \mu_{21} + \mu_{03}}{a}$$

$$\epsilon_{32} = x_g + \frac{x_g\mu_{20} + y_g\mu_{11} + \mu_{12} + \mu_{30}}{a}$$

The variables $c_{i_{wx}}$, $c_{i_{wy}}$, $c_{j_{wx}}$ and $c_{j_{wy}}$ are obtained in [18], while α_{wx} and α_{wy} are described in [16].

3   Methodology

3.1   Coral Detector

The coral detector takes an input image from the database of training images,and the next stages in Fig. 2 are explained as follows in [7]:

1. Pre-processing: We convert the RGB image to grayscale, and we normalise the pixel values.
2. Feature extraction: We extract texture feature descriptors by convolution using a bank of Gabor Wavelet filters. Thus, each feature vector is associated to every pixel of the image.
3. Discrimination: We use machine learning, the Decision Trees algorithm, for pixel classification between coral and non-coral reef.
4. Post-processing: We remove false positives and false negatives by selecting the detections with the biggest contours.

3.2   Visual Servoing Algorithm

This section describes the steps that implement the velocity controller of the Coral-bot. We develop an eye-in-hand visual servoing whose camera is located at the robot arm. A set of point features and moment features are generated to control 4 *dof* of an autonomous underwater robot. We describe the stages to be developed in this application as follows:

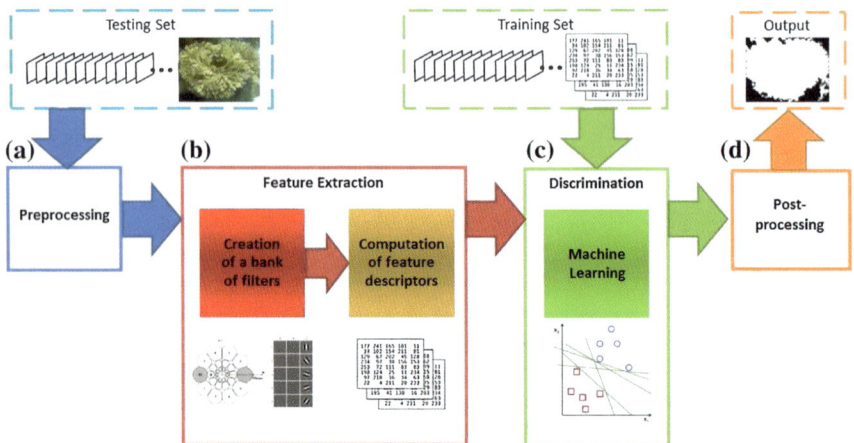

Fig. 2   Block diagram of coral reef detector: **a** Pre-processing, **b** Feature extraction, **c** Discrimination and **d** Post-processing

1. We select the biggest detection that fits a rectangle. Four vertices of this polygon represent the input features of the system.
2. We estimate depth, Z, of these features by computing the determinant of the Homography matrix H [19] that results of using the four correspondent points with the desired point features and the desired depth Z^d; as follows in Eq. (9)

$$Z = Z^d det(H) \qquad (9)$$

3. We compute the velocity screw ξ by applying Eq. (3) and substituting the interaction matrices: Eq. (4) for point features, and Eq. (8) for moment features.
4. We transform the velocity screw in the camera frame $^c v = \xi$ to the velocity in the vehicle frame by applying the velocity twist matrix $^v V_c$ [20] in (10)

$$^v v = {}^v V_c {}^c v \qquad (10)$$

5. We apply a PID compensator [21] in order to improve the stability of the system, which is sensitive to noise that comes from missing detections, or disturbances introduced by the ocean environment.

4 Results

4.1 Coral Detector

The experiments of this work consisted of evaluating nine machine learning algorithms: Decision Trees (DTR), Random Forest (RTR), Extremely Randomised Trees (ERT), Boosting (BOO), Gradient Boosted Trees (GBT), Normal Bayes Classifier (NBA), Expectation Maximisation (EMA), Neural Networks (MLP), Support Vector Machines (SVM); which are available in the OpenCV library. We used a database of 621 images of coral reef from Belize with 110 images for training and 511 images for testing.

Experts selected regions of interest of 110 images of training to generate 6.800.071 feature vectors. These features are used for training the aforementioned machine learning algorithms. Figure 3 describes the results of comparing their testing times and accuracies.

The results make evident the fastest machine learning algorithm: the Decision Trees. It classifies an image of 1024×768 pixels in 70 ms. The time of the rest of the algorithms are described in Fig. 3a.

The accuracy provides the correct prediction rate over the number of total evaluated cases. Although ERT, MLP, GBT and BOO obtain more than 70 % of accuracy (see Fig. 3b), they do not reveal visually a good performance. DTR, NBA and EMA develop a better discrimination with accuracies around 60 %. SVM produces a poor performance among the algorithms.

(a)

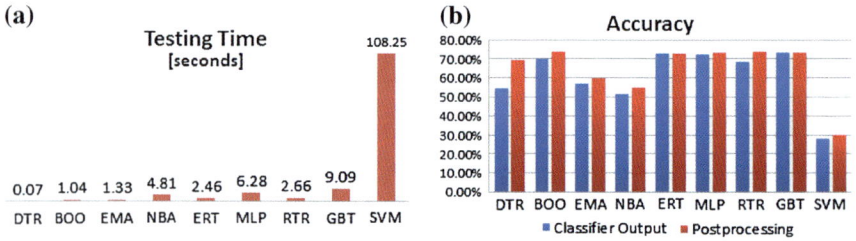

(b)

Fig. 3 Metrics for evaluation. **a** Running time. **b** Accuracy

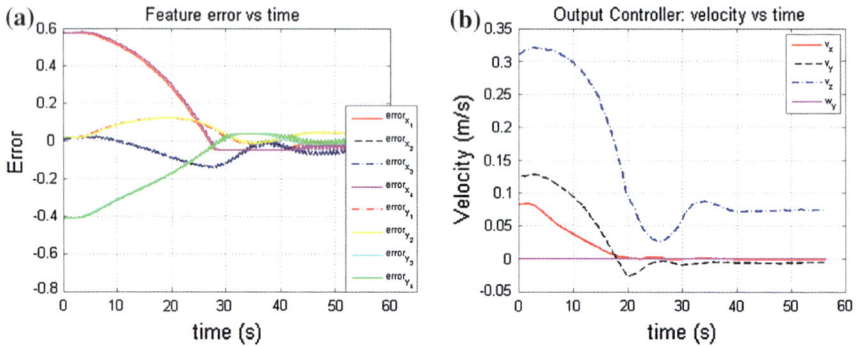

Fig. 4 Error and velocity curves using point features. **a** Error. **b** Velocity

4.2 Visual Servoing

The performance of these two visual servoing approaches (point and moment features) is compared by using a pure translation of a station keeping task. These experiments were performed using the underwater simulator, UWSim.

The error trajectories of the point features and the robot velocities are observed in Fig. 4a. These errors converge continuously to zero, with oscillations that generate an under-damped behaviour. The action of the PID compensator plays an important role by attenuating disturbances. Figure 4b shows the compensated velocities of the Coralbot. We notice a behaviour of deceleration in v_z before the first 20 s. After that interval, the vehicle has to move up to reach the desired features. It generates a cycle of oscillation that is compensated in less than 10 s.

The velocities with moment features, are shown in Fig. 5b. A significant undershoot demonstrates the difficulties to control heave motion by using the target area. This feature is very sensitive to noise that comes from detector due to the target shape changes slightly all the time.

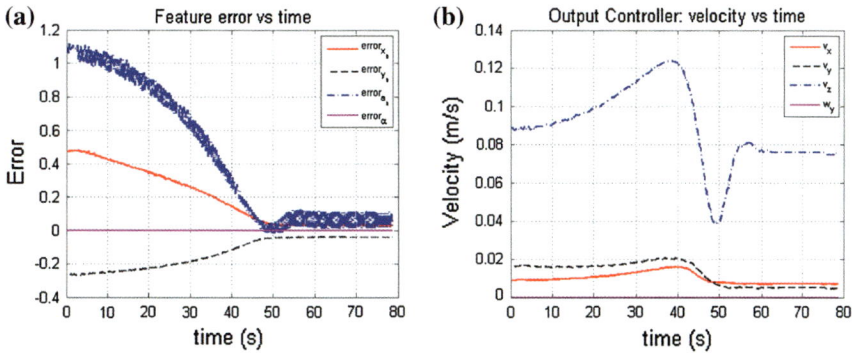

Fig. 5 Error and velocity curves using moment features. **a** Error. **b** Velocity

5   Conclusions

This article presents the results of developing two visual servoing algorithms: an Image-based approach and a Moment-based approach. The first approach uses points as features, while the second approach introduces moment features. Point features offer a good system performance for developing tasks of pure translation and station keeping. PID compensator rejects the disturbances generated by the coral detector and improves the results. Moment features represent a good alternative for implementing control strategies due to each moment feature is associated with a velocity direction. We design the control action assuming a proper decoupling of the system.

We implement a coral detector that classifies among two classes: coral an non-coral reef. The discrimination process is based on supervised machine learning. We classify texture feature descriptors by using a bank of Gabor Wavelet filters. Nine classifiers were evaluated using running time and accuracy. It is difficult to identify which classifier outperforms other algorithms. However, the classifiers such as Extremely Randomised Trees, Gradient Boosted Trees, Neural Networks, and Support Vector Machines; show difficulties to discriminate coral and non-coral classes. In contrast, Decision Trees, Boosting, Expectation Maximisation, Normal Bayes Classifier and Random Forest present promising visual classification.

For this reason, we recommend the use of an Image-based approach for visual servoing and Decision Trees algorithm for coral detections, in order to implement proper speed controllers in the Coralbot.

Acknowledgments This work was developed at the Ocean Systems Lab and the Computer Vision Lab in Heriot-Watt University, thanks to the support of the European Commission and the VIBOT Consortium.

References

1. Coralbot Team—About coral reefs and current approaches for restoration. http://www.coralbots.org/.
2. Stough, J. V., Greer, L., & Benson, M. (2012). Texture and color distribution-based classification for live coral detection. In *12th International Coral Reef Symposium*.
3. Marcos, M. S. A., David, L., Peñaflor, E., Ticzon, V., & Soriano, M. (2008). Automated benthic counting of living and non-living components in Ngedarrak Reef, Palau via subsurface underwater video. *Journal of Environmental Monitoring and Assessment, 145*, 177–184.
4. Beijbom, O., Edmunds, P. J., Kline, D. I., Mitchell, B. G., & Kriegman, D. (2012). Automated annotation of coral reef survey images. In *2012 IEEE Conference on Computer Vision and Pattern Recognition (CVPR)* (pp. 1170–1177). Rhode Island.
5. Johnson-Roberson, M., Kumar, S., & Willams, S. (2006). Segmentation and classification of coral for oceanographic surveys: A semi-supervised machine learning approach. In *OCEANS 2006—Asia Pacific*. Singapore: IEEE Press.
6. Purser, A., Bergmann, M., Lundälv, T., Ontrup, J., & Nattkemper, T. W. (2009). Use of machine-learning algorithms for the automated detection of cold-water coral habitats: A pilot study. *Journal of Marine Ecology Progress Series, 397*, 241–251.
7. Tusa, E., Reynolds, A., Lane, D. M., Robertson, N. M., Villegas, H., & Bosnjak, A. (2014). Implementation of a fast coral detector using a supervised machine learning and Gabor Wavelet feature descriptors. *Sensor Systems for a Changing Ocean (SSCO), 2014 IEEE, Brest*, (pp. 1–6). doi: 10.1109/SSCO.2014.7000371
8. Mitchell, T. M. (1997). *Machine learning*. New York: McGraw-Hill.
9. Loh, W. Y. (2011). Classification and regression trees. *Journal of WIREs Data Mining and Knowledge Discovery, 1*, 14–23.
10. Bradski, G., & Kaehler, A. (2008). *Learning OpenCV: Computer vision with the OpenCV library*. Cambridge: OReilly.
11. Geurts, P., Ernst, D., & Wehenkel, L. (2006). Extremely randomized trees. *Journal of Machine Learning, 63*, 3–42.
12. Luber, M., Spinello, L., & Arras, K. O. (2011). People tracking in rgb-d data with on-line boosted target models. In *International Conference on Intelligent Robots and Systems (IROS)*. San Francisco: IEEE Press.
13. Friedman, J. H. (2000). Greedy function approximation: A gradient boosting machine. *Annals of Statistics, 29*, 1189–1232.
14. Bilmes, J. A. (1998). A gentle tutorial of the EM algorithm and its application to parameter estimation for Gaussian mixture and hidden Markov models, Technical Report. International Computer Science Institute.
15. Chaumette, F., & Hutchinson, S. (2006). Visual servo control, Part I: Basic approaches. *IEEE Robotics and Automation Magazine, 13*, 82–90.
16. Tahri, O., & Chaumette, F. (2005). Point-based and region-based image moments for visual servoing of planar objects. *IEEE Transactions on Robotics, 21*, 1116–1127.
17. Chaumette, F. (2004). Image moments: A general and useful set of features for visual servoing. *IEEE Transactions on Robotics, 20*, 713–723.
18. Tahri, O. (2004). *Application of moment to visual servoing and pose estimation*. Ph.D. thesis, Rennes.
19. Lots, J., Lane, D., Trucco, E., & Chaumette, F. (2001). A 2-D visual servoing for underwater vehicle station keeping. In *2001 IEEE International Conference on Robotics and Automation* (Vols. I–IV, pp. 2767–2772). Seoul: IEEE Press.
20. Bakthavatchalam, M., Chaumette, F., Marchand, E., Novotny, F., Saunier, A., Spindler, F., et al. (2012). Geometric transformations and objects. INRIA. http://www.irisa.fr/lagadic/visp/documentation/visp-tutorial-geometric-objects.pdf.
21. Zhong, J. (2006). PID controller tuning: A short tutorial. http://wwwdsa.uqac.ca/~rbeguena/Systemes_Asservis/PID.pdf.

Underwater Vehicle Modeling and Control Application to Ciscrea Robot

Rui Yang, Irvin Probst, Ali Mansour, Ming Li and Benoit Clement

Abstract Underwater competitions confirm that the PID yaw controller is less effi-cient for low mass Autonomous Underwater Vehicle (AUV) to handle the robot uncertainties. Nonlinear hydrodynamic behavior, waves, current, AUV bouyance change, motor calibration variations, sensor disturbance and battery variations per-turbate the PID control behavior a lot. Therefore, in this paper we present a model based robust controller to control the yaw heading of AUV CISCREA. The modeling result was verified with experiments, and the robust controller was simulated.

Keywords Underwater vehicle · Added mass · Damping · Robust

1 Introduction

AUVs are playing important roles in underwater activities. For some applications: undersea pipeline survey, infrastructure inspections and large vehicle wet main-tenance tasks, a small size cubic AUV is preferred. Indeed, small AUVs can be deployed to explore areas where HOVs (Human Occupied Vehicles) and ROVs

R. Yang · M. Li
College of Engineering, Ocean University of China, Qingdao 266100, China
e-mail: yang.rui@ensta-bretagne.fr

M. Li
e-mail: limingneu@ouc.edu.cn

R. Yang · I. Probst · A. Mansour · B. Clement (✉)
ENSTA Bretagne, Brest Cedex 9, 29806 Brest, France
e-mail: benoit.clement@ensta-bretagne.fr

I. Probst
e-mail: irvin.probst@ensta-bretagne.fr

A. Mansour
e-mail: mansour@ieee.org

R. Yang · A. Mansour · B. Clement
Lab-STICC(UMR CNRS 6285), Technopole Brest Iroise, Brest Cedex 3,
29238 Brest, France

© Springer International Publishing Switzerland 2016   89
B. Zerr et al. (eds.), *Quantitative Monitoring of the Underwater Environment*,
Ocean Engineering & Oceanography 6, DOI 10.1007/978-3-319-32107-3_9

(Remote Operating Vehicles) are limited to operate. Meanwhile, the cubic AUVs show more degrees of freedom than torpedo-shaped AUVs in motion. Especially, they can hover and enter complex underwater spaces.

Achieving maneuverability of small AUV depends on two key factors: an accurate hydrodynamic model and an advanced control system. In [24], Yamamoto pointed out that a model-based control system is more effective if the vehicle's dynamics are modeled to some extent. Meanwhile, in [9], Ferreira et al. showed that an empirical linear model often fails to represent the dynamics of the AUV over a wide operating region. Therefore, obtaining hydrodynamic models of the complex-shaped cubic AUV is one of the key points for better maneuverability.

Actually, many methods exist to model ocean vehicles, including scaled experiments, full-scale experiments, empirical formula approximations and computational dynamic approaches. Scaled and full-scale experiments are capable to provide accurate hydrodynamic models. However, they usually need expensive devices, such as towing tanks. Besides, most of the time experimental modeling results are not control-oriented. Experimental methods without towing tanks also exist as presented in [1, 19], and free decay approach is presented by Ross in [21]. Empirical formula approximation is well proved on torpedo-shaped AUVs, usually slender bodies, as mentioned in [10, 19]. Nonetheless, empirical formula method requires deeper knowledge and experiences to simplify the AUV into elementary components. Especially, for a complex-shaped AUV, the simplification is too complicated. Computational approaches use potential theory and finite element theory based CFD (Computational Fluid Dynamic) software such as WAMIT™, MCC (Marine Craft Characteristics, this freeware is created by ENSTA Bretagne which can be downloaded from [16]), SHIPMO™, ANSYS-CFX™, ANSYS-FLUNT™, ANSYS-AQWA™, STAR-CCM+™ and SeaFEM™. They are capable to predict hydrodynamic parameters for a complex-shaped AUV with very low cost. In [6], it is shown the efficiency of WAMIT™ to predict the added mass matrix. In [5] ANSYS-CFX™ was employed for AUV damping analysis. In [25], numerical model of CISCREA AUV was derived and verified.

Regardless of modeling methods, the value of a hydrodynamic model depends on how robust your control scheme can adopt it. AUVs are generally designed to operate in the ocean environments. Therefore, numerous uncertainties are encountered, including parameter variations, non-linear hydrodynamic effects, sensor measurement delays and ocean current disturbances. Owing to these unpredictable problems, traditional control methods, such as PID (Proportional Integral Derivative) and LQG (Linear Quadratic Gaussian), are not considered as efficient to guarantee both stability and high performance, see [4]. In SAUC-E [22] and euRathlon [7] competition we found PID yaw controller is less efficient for low mass AUV. Consequently, advanced control algorithms should be involved, such as the adaptive control scheme in [15], robust control scheme in [2] and interval approach in [14]. Note that robust control schemes were shown successfully in [8, 20] for torpedo-shaped AUVs.

In this works, we appointed the semi-AUV CISCREA, as shown in Fig. 1a, and characterized in Table 1.

(a)

(b)

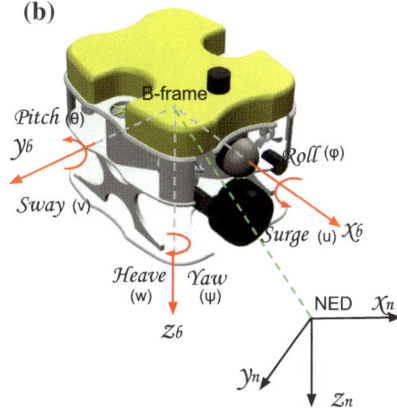

Fig. 1 CISCREA robot, **a** view in water **b** frame notions of underwater vehicles

Table 1 CISCREA characteristics

Size	0.525 m (L) 0.406 m (W) 0.395 m (H)
Weight in air	15.56 kg (without payload and floats)
Degrees of freedom	Surge, sway, heave and yaw
Propulsion	2 vertical and 4 horizontal propellers
Speed	2 knots (Surge) &1 knot (Sway, Heave)
Depth rating	50 m
On-board battery	2–4 h

This paper is organized in five sections; main notions for the underwater vehicle are presented in Sect. 2; the computational modeling approach and numerical results are given in Sect. 3; experiments to verify the numerical hydrodynamic model is described in Sect. 4; Sect. 5 demonstrates a robust control framework with Matlab simulation; and finally conclusions are drawn in Sect. 6.

2 AUV Modeling Framework

For marine systems, usually two coordinate systems, NED-frame (North East Down) and B-frame (Body fixed reference) are introduced for convenience as presented by Fossen in [12] and shown in Fig. 1b.

In this section, CISCREA dynamics are represented by the marine vehicle formulation proposed by Fossen in [11, 12], and the SNAME [23] (Society of Naval Architects and Marine Engineers) notions in [23]. Positions, angles, linear and angu-

Table 2 The notation of SNAME for marine vessels

	Positions & angles	Linear & angular velocities	Forces & moments
Coordinate	NED-frame	B-frame	B-frame
Surge	x	u	X
Sway	y	v	Y
Heave	z	w	Z
Roll	ϕ	p	K
Pitch	θ	q	M
Yaw	ψ	r	N

lar velocities, force and moment definitions are reflected in Table 2. The position vector η, velocity vector v and force vector τ are defined as:

$$\eta = [x, y, z, \phi, \theta, \psi]^T; \; v = [u, v, w, p, q, r]^T \; ; \; \tau = [X, Y, Z, K, M, N]^T$$

It is known that kinematic relation of velocity vector v and position vector η is expressed as follows [12]:

$$v = J(\Theta)\dot{\eta} \tag{1}$$

where, $J(\Theta) \in \mathbb{R}^{6\times6}$, stands for a transformation matrix between B-frame and NED-frame, $\Theta = [\phi, \theta, \psi]^T$, $c(\cdot) = cos(\cdot), s(\cdot) = sin(\cdot)$ and $t(\cdot) = tan(\cdot)$.

$$J(\Theta) = \begin{bmatrix} R(\Theta) & \mathbf{0}_{3\times3} \\ \mathbf{0}_{3\times3} & T(\Theta) \end{bmatrix}, \qquad T(\Theta) = \begin{bmatrix} 1 & s(\psi)t(\theta) & c(\phi)t(\theta) \\ 0 & c(\phi) & s(\phi) \\ 0 & \frac{s(\phi)}{c(\theta)} & \frac{c(\phi)}{c(\theta)} \end{bmatrix}$$

$$R(\Theta) = \begin{bmatrix} c(\psi c(\theta) & -s(\psi)c(\phi) + c(\psi)s(\theta)s(\phi) & s(\psi)s(\phi) + c(\psi)c(\phi)s(\theta) \\ s(\psi c(\theta) & c(\psi)c(\phi) + s(\phi)s(\theta)s(\psi) & -c(\psi)s(\phi) + s(\theta)s(\psi)c(\phi) \\ -s(\theta) & c(\theta)s(\phi) & c(\theta)c(\phi) \end{bmatrix}$$

Depending on [12], rigid-body hydrodynamic forces and moments can be linearly superimposed. Therefore, the overall non-linear underwater model can be characterized by two parts, the rigid-body dynamic (2) and hydrodynamic formulations (3) (hydrostatics included). Table 3 shows parameter definitions for this model.

$$M_{RB}\dot{v} + C_{RB}(v)v = \tau_{env} + \tau_{hydro} + \tau_{pro} \tag{2}$$

$$\tau_{hydro} = -M_A\dot{v} - C_A(v)v - D(|v|)v - g(\eta) \tag{3}$$

Rigid-body mass inertia matrix M_{RB} is defined in Eq. (4), where m is the mass and $r_G = [x_G, y_G, z_G]^T$ is the vector from O_b (origin of B-frame) to CG (center of gravity). If $r_G = 0$, i.e., $O_b \equiv CG$, then the matrix M_{RB} will be simplified. Moreover, symmetric properties of cubic AUV, in $x = 0$ and $y = 0$ planes, can be used to simplify the inertia components to a rough diagonal form.

Table 3 Nomenclature of the notations

Parameter	Description
M_{RB}	AUV rigid-body mass and inertia matrix
M_A	Added mass matrix
C_{RB}	Rigid-body induced coriolis-centripetal matrix
C_A	Added mass induced coriolis-centripetal matrix
$D(\lvert v \rvert)$	Damping matrix
$g(\eta)$	Restoring forces and moments vector
τ_{env}	Environmental disturbances (wind, waves and currents)
τ_{hydro}	Vector of hydrodynamic forces and moments
τ_{pro}	Propeller forces and moments vector

$$M_{RB} = \begin{bmatrix} m & 0 & 0 & 0 & mz_G & -my_G \\ 0 & m & 0 & -mz_G & 0 & mx_G \\ 0 & 0 & m & my_G & -mx_G & 0 \\ 0 & -mz_G & my_G & I_x & -I_{xy} & -I_{xz} \\ mz_G & 0 & -mx_G & -I_{yx} & I_y & -I_{yz} \\ -my_G & mx_G & 0 & -I_{zx} & -I_{zy} & I_z \end{bmatrix} \tag{4}$$

C_{RB} and C_A contribute to the centrifugal force. A practical way to calculate these two matrices using M_{RB}, M_A, v and the matlab function "m2c.m" is introduced in MSS (Marine System Simulator) in [17]. In our case, these two matrices are neglected, because the vehicle speed is low enough to be considered, $C(v) \approx 0$.

For an AUV with neutral buoyancy, the weight W is approximately equal to the buoyancy force B. In Eq. (5), g is the gravity acceleration, ρ is the fluid density, and ∇ is the displaced fluid volume.

$$W = mg, B = \rho g \nabla \tag{5}$$

As pointed out by [12], the restoring forces and moments vector $g(\eta)$ can be simplified as in (6), where $BG = [BG_x, BG_y, BG_z]^T$ is the distance from the CG to CB (the buoyancy center).

$$g(\eta) = \begin{bmatrix} 0 \\ 0 \\ 0 \\ -BG_y W cos(\theta) cos(\phi) + BG_z W cos(\theta) sin(\phi) \\ -BG_z W sin(\theta) + BG_x W cos(\theta) sin(\phi) \\ -BG_x W cos(\theta) sin(\phi) - BG_y W sin(\theta) \end{bmatrix} \tag{6}$$

For CISCREA, CB and CG can be located using trial and error method on adding and removing the payload and floats.

The marine disturbances, such that the wind, waves and current contribute to the environmental effect τ_{env}. But for an underwater vehicle, only current is considered since wind and waves have negligible effects on an AUV during underwater operations.

In order to put forward the model in the same coordinate, a transformation is made according to Eqs. (1), (2) and (3). Cubic AUV model under both NED-frame and B-frame is transformed into NED-frame as follows:

$$M^*\ddot{\eta} + D^*(|v|)(\dot{\eta}) + g^*(\eta) = \tau_{pro} + \tau_{env} \qquad (7)$$

$$M^* = J^{-T}(\Theta)(M_{RB} + M_A)J^{-1}(\Theta);$$

$$D^*(|v|) = J^{-T}D(|v|)J^{-1}(\Theta);$$

$$g^*(\eta) = J^{-T}g(\eta)$$

Two hydrodynamic parameters added mass, $M_A \in \mathbb{R}^{6\times6}$, and damping, $D(|v|) \in \mathbb{R}^{6\times6}$, should be carefully involved in the AUV model. Added mass is a virtual conception representing the hydrodynamic forces and moments. Any accelerating emerged-object would encounter this M_A due to the inertia of the fluid. For a cubic AUV, added mass in some directions are generally larger than the rigid-body mass [25]. Damping in the fluid consists of four parts, as described in Eq. (8): Potential damping $D_P(|v|)$, skin friction $D_S(|v|)$, wave drift damping $D_W(|v|)$ and vortex shedding damping $D_M(|v|)$.

$$D(|v|) = D_P(|v|) + D_S(|v|) + D_W(|v|) + D_M(|v|) \qquad (8)$$

Details of the two hydrodynamic parameters are discussed in the following section.

3   Computational Solutions for Dynamic and Hydrodynamic Parameters

This section is dedicated to calculate numerically dynamic and hydrodynamic parameters: Mass inertia matrix M_{RB}, added mass matrix M_A and damping matrix $D(|v|)$. Due to the complex structure, traditional empirical formula approximation is not as efficient as for slender bodies [19]. To solve this problem, we propose to calculate hydrodynamic models using CFD software as follows:

- Mass inertia matrix M_{RB} is calculated using PRO/ENGINEER[TM].
- Added mass matrix M_A is calculated using radiation/diffraction program MCC and WAMIT[TM].
- Hydrodynamic programs ANSYS-CFX[TM] and STAR-CCM+[TM] are studied to predict damping behavior $D(|v|)$.

3.1 Rigid-Body Mass Inertia Matrix

Due to different density components, the inertia parameters and CG of CISCREA are hard to be calculated according to [12] (Eq. (9)).

$$m = \int_V \rho_m dV, \quad I = \int_V r^2 \rho_m dV \tag{9}$$

Here ρ_m is the density of volume element dV, V is the volume of the body, r is the distance between volume element dV and CG.

A practical way is to measure the principal components of the AUV and calculate M_{RB} automatically with CAD software PRO/ENGINEER™. This process is shown in Fig. 2.

The output of PRO/ENGINEER™ for CISCREA around CG (O_b), M_{RB}, is listed in Eq. (10) (kg and kg/m^2).

$$M_{RB} = \begin{bmatrix} 15.643 & 0 & 0 & 0 & 0 & 0 \\ 0 & 15.643 & 0 & 0 & 0 & 0 \\ 0 & 0 & 15.643 & 0 & 0 & 0 \\ 0 & 0 & 0 & 0.2473 & 0 & 0.0029 \\ 0 & 0 & 0 & 0 & 0.3698 & 0 \\ 0 & 0 & 0 & 0.0029 & 0 & 0.3578 \end{bmatrix} \tag{10}$$

By neglecting small off-diagonal inertia elements, the above matrix can be simplified to a diagonal matrix:

$$M_{RB} = diag\left(\begin{bmatrix} 15.643 & 15.643 & 15.643 & 0.2473 & 0.3698 & 0.3578 \end{bmatrix}\right) \tag{11}$$

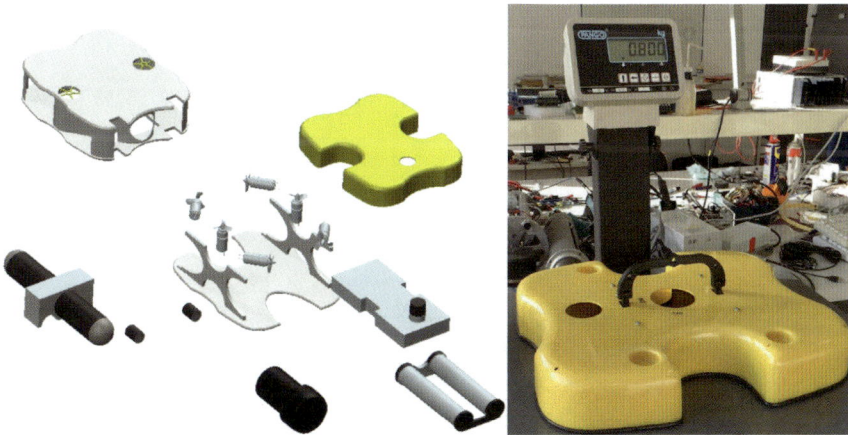

Fig. 2 Measure and calculate mass inertia matrix in PRO/ENGINEER™

Table 4 Sphere added mass (Radius 1 m, density 1 kg/m³, depth 10 m, 1024 surfaces)

	Theoretical	WAMIT	MCC
Surge(kg)	2.0944	2.084236	2.106

3.2 Added Mass Matrix

It is mentioned in [18], variations of underwater vehicle geometry play indeed a role on the added mass of the AUV. However, empirical formulas predicting is inaccurate for complex-shaped AUV. To solve this issue, boundary element method based WAMIT™ and MCC are used [25].

Knowing that the theoretical sphere added mass is given by $2/3\pi\rho r^3$. Therefore, we use sphere to verify configurations. Results are compared in Table 4. Import the same input control files of the sphere calculation with CISCREA geometry file to WAMIT™ and MCC, added mass matrix results are calculated and compared.

WAMIT™ output is listed in M_{A1} (12), and MCC in M_{A2} (13) (Mass: kg, Inertia: kg m²). Note that the vehicle speed is low enough to neglect the small off-diagonal elements.

$$M_{A1} = \begin{bmatrix} 11.985 & -0.091 & -0.105 & 0.039 & 0.308 & 0.012 \\ 0.149 & 20.261 & -0.147 & 0.085 & -0.013 & -0.758 \\ 0.111 & -0.129 & 67.141 & -0.033 & 2.530 & 0.064 \\ 0.122 & 0.319 & -0.056 & 0.385 & 0.003 & -0.011 \\ 0.407 & -0.001 & 2.543 & -0.002 & 0.791 & 0.002 \\ -0.003 & -0.758 & 0.064 & -0.003 & 0.004 & 0.138 \end{bmatrix} \quad (12)$$

$$M_{A2} = \begin{bmatrix} 11.8 & 4.08 & 9.41 & 0.326 & 0.349 & -0.267 \\ 4.53 & 17.9 & -10.3 & 0.492 & -0.913 & 0.233 \\ 8.6 & -12.3 & 52.7 & -2.88 & -7.94 & 1.49 \\ 0.256 & 0.676 & -2.74 & 0.91 & 0.573 & 0.0087 \\ -0.067 & -0.628 & -9.17 & 0.655 & 1.54 & 0.04 \\ -0.184 & 0.162 & 1.29 & -0.0289 & -0.0252 & 0.0854 \end{bmatrix} \quad (13)$$

3.3 Damping Matrix

As mentioned before in Eq. (8), four elements contribute to $D(|v|)$, the damping matrix of marine vehicles. For CISCREA, potential damping $D_P(|v|)$ is negligible comparing to other terms [12], waves drift damping $D_W(|v|)$ is also negligible, and waves are assumed to act on surface vehicles. Therefore, skin friction damping $D_S(|v|)$ and vortex shedding damping $D_M(|v|)$ are the only parameters left to study.

As presented in [12], usually damping terms contribute to linear and quadratic damping. The Eq. (14) is introduced,

$$D(|v|) = D + D_n(|v|) \tag{14}$$

where D is the linear damping matrix and $D_n(|v|)$ is a quadratic damping matrix. If vehicle velocities are sufficiently high the D can be generally neglected. Otherwise, $D_n(|v|)$ is negligible.

Empirical formulas are used to roughly determine the damping behaviors. The quadratic damping $f(U)$ force is represented as follows [3]:

$$f(U) = -\frac{1}{2}\rho_d C_D(R_n)AU|U| \tag{15}$$

where U is the vehicle velocity, ρ_d is the fluid density, A is the cross-sectional area projected on the fluid and $C_D(R_n)$ is the damping coefficient, which is a function of Reynolds number R_n.

$$R_n = \frac{\rho_d U D_{CL}}{v_{is}} \tag{16}$$

For R_n in (16), v_{is} is the fluid viscosity and D_{CL} is the characteristic length. In our study, the fluid is chosen to be the seawater as described in [12]. CISCREA generally operates at a speed U range from 0 to 1 m/s. Robot characteristic length is approximately 0.5 m. In this case, the R_n of CISCREA is generally around 10^8 to 10^9. This implies that the damping of CISCREA is not in the critical area, i.e., from 10^5 to 10^6 between the laminar and the turbulent flow. Therefore, a converged constant damping coefficient is encountered, and nonlinear quadratic damping behavior is expected. The variations of C_D chart with respect to R_n can be found in [3].

Due to the complex-shaped geometry of CISCREA, the empirical formulas are impractical to predict the damping effects. Meanwhile, it is hard to tell which part of the linear and quadratic damping play the major role for CISCREA. So, finite element theory based CFD software ANSYS-CFX[TM] and STAR-CCM+[TM] are used to estimate the relationship among damping forces, damping moments and vehicle velocities, angular velocities.

AUV is fixed in rectangular (surge, sway, heave) or cylindrical (yaw) water tanks, as shown in Fig. 3. Fluid moves in the tank with the speed variations from 0 to 0.5 m/s for translational motion and 0 to 5 rad/s for rotational motion. Respectively, speed growth interval of 0.1 m/s and 0.5 rad/s are selected.

The configurations used in CFX and STAR-CCM+ are listed in Table 5, and the damping force (moment) results are shown in Table 6.

Second order polynomial lines are implemented to approximate the relationship between damping and velocities, see Figs. 4, 7d and Table 7. Our results show that quadratic damping dominates the damping effects.

(a)   **(b)**

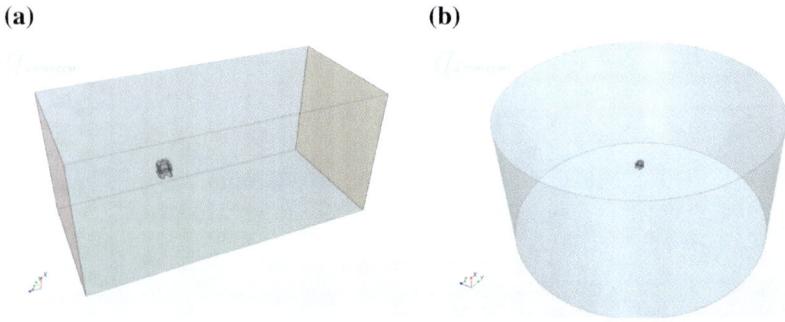

Fig. 3 CISCREA set up in translational and rotational water tanks. **a** Sway in STAR-CCM+.
b Yaw rotation in STAR-CCM+

Table 5 Configurations of CFX and STAR-CCM+

Parameter	CFX Configuration	STAR-CCM+ Configuration
Tank	10.5 m (L) 4.5 m (W) 4.5 m (H)	9 m (L) 4 m (W) 4 m (H)
Cylinder	–	8 m(H) 8 m(radius)
Fluid	Steady	Steady
Density	$\rho = 1023\,\mathrm{kg/m^3}$	$\rho = 1023\,\mathrm{kg/m^3}$
Viscosity	$1.56 \times 10^{-6}\,\mathrm{kg/(s \cdot m)}$	$1.56 \times 10^{-6}\,\mathrm{kg/(s \cdot m)}$
Turbulence	1 % at inlet boundary, $(k - \omega)$	1 % at inlet boundary, $(k - \omega)$
Mesh	588221 elements for heave (around for surge and sway)	1002637 cells, 3002121 faces for sway (around for surge and heave)
Convergence	10^{-4}	100 steps $(<10^{-5})$
Roughness	PVC 0.0015–0.007 (mm)	Wall

Table 6 Damping forces and moments at different velocities

	0.1 m/s	0.2 m/s	0.3 m/s	0.4 m/s	0.5 m/s			
CFX								
Surge (N)	0.287	1.146	2.577	4.579	7.222			
Sway (N)	0.537	2.14	4.815	8.561	13.382			
Heave (N)	0.51	3.319	7.47	13.28	20.751			
CCM+								
Surge (N)	0.273	1.06	2.39	4.253	6.539			
Sway (N)	0.5077	2.011	4.4531	8.0222	12.2759			
Heave (N)	0.8393	3.298	7.43	12.974	20.745			
CCM+ (rad/s)	0.5	1	1.5	2	2.5	3	3.5	4
Yaw (N · m)	0.038	0.149	0.338	0.593	0.932	1.33	1.792	2.381

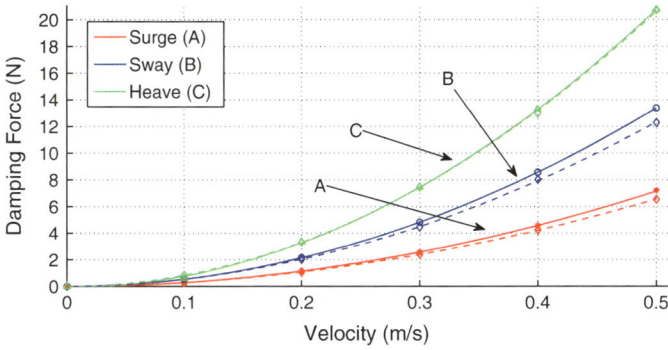

Fig. 4 Damping force and velocity (CFX: *solid line*, STAR-CCM+: *dash line*)

Table 7 CFD results curve fitting

	CFX & RMSE		STAR-CCM+ & RMSE	
Surge	$y = 28.6x^2 + 0.0089x$	0.01773	$y = 25.75x^2 + 0.2406x$	0.02294
Sway	$y = 53.52x^2$	0.00188	$y = 48.39x^2 + 0.4512x$	0.0595
Heave	$y = 83.42x^2$	0.103	$y = 82.44x^2$	0.1144
Yaw	–	–	$y = 0.1479x^2 + 0.001328x$	0.009881

4 Experimental Model Results

In order to verify the efficiency of the numerical model obtained in Sect. 3. Real world experiments have been conducted on the open-loop CISCREA to verify the translational and rotational damping parameters.

Bollard thrusts of propellers are measured, see Fig. 5. During experiment, CIS-CREA is driving in the surge and sway directions inside a pool (4 m × 4 m × 3.5 m) from one end to another, and dive in the heave direction from top to bottom. The process of yaw rotation drives CISCREA spin in the middle of the pool until it reaches a constant angular velocity. Meanwhile, all these moving processes were captured by a 15 fps CCD camera on top and another camera with 25 fps underwater, as shown in Fig. 6.

The propelling force, the pool size and the time hitting the wall or single rotational lap are known, we can build damping and velocity relationship based on this information. First of all, we can verify that giving any propelling input, the final velocity of AUV should converge to a steady constant speed. We simply assume the surge dynamic is a linear equation (17), where D_L is the unknown linear damping coefficient, x is the surge position of CISCREA. As a result, the convergence indicates that the average speed can be measured after a specific time. Position and camera frame information can provide this average speed for surge, sway and yaw motions. For the heave motion, we use the final speed of the linear model (17).

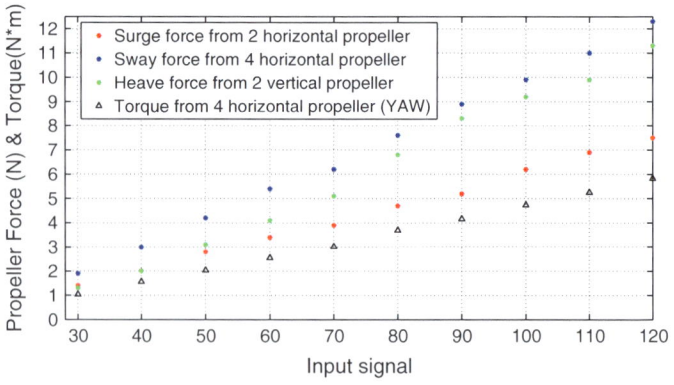

Fig. 5 Forces and torques measurements

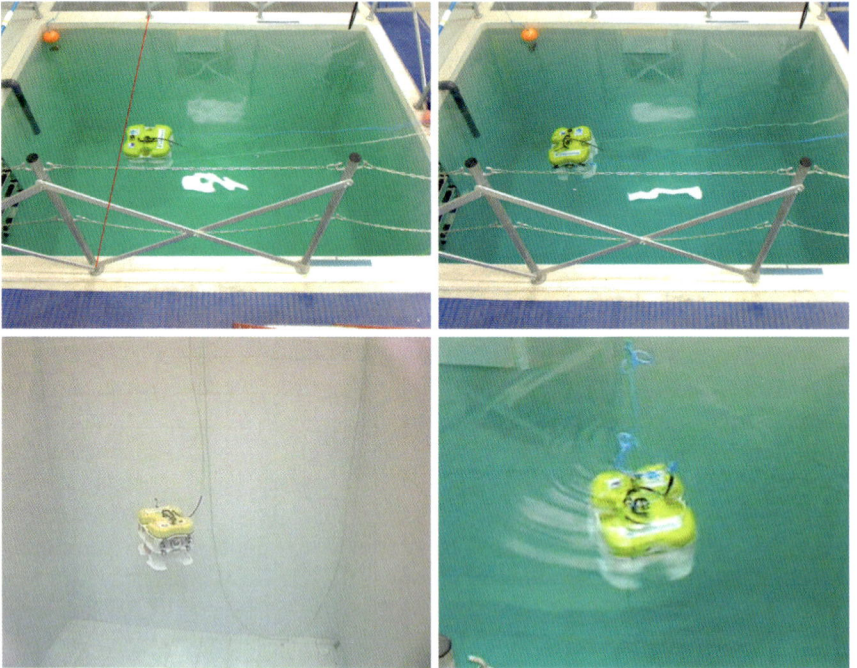

Fig. 6 CISCREA moving in surge, sway, heave and yaw directions

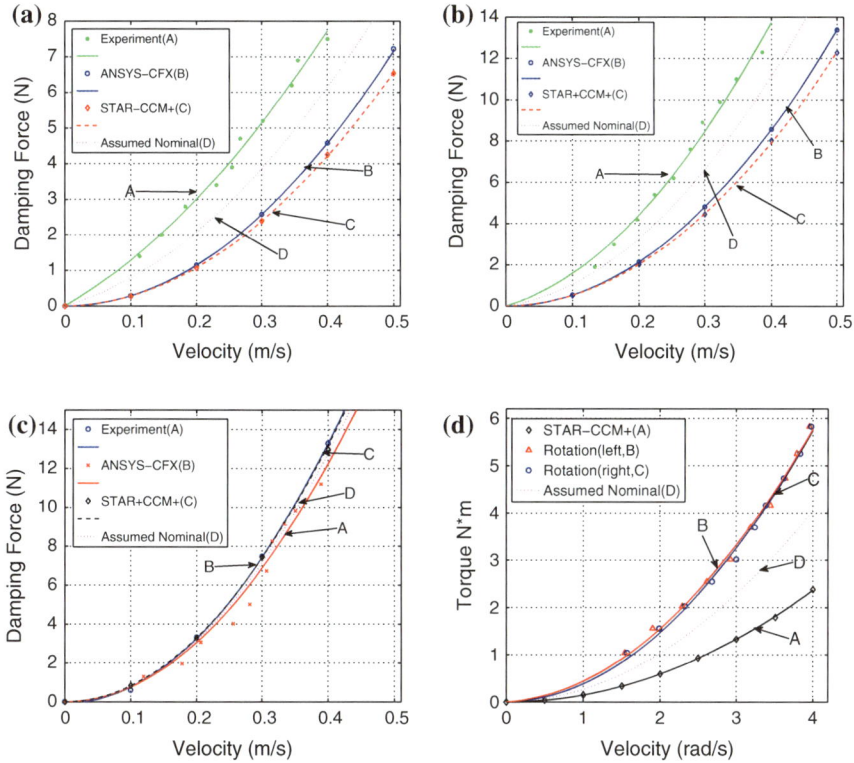

Fig. 7 Comparison of ANSYS and experiment damping and velocities. **a** Surge. **b** Sway. **c** Heave. **d** Yaw

$$(M_{RB} + M_A)\ddot{x} + D_L\dot{x} = \tau_m \qquad (17)$$

Finally, experimental damping and velocity results and polynomial lines are compared to CFD results in Fig. 7 and Table 8. The gap in Fig. 7a, b are mainly caused by the drag of cables and ropes, which are playing opposite efforts, as shown in Fig. 7c. In additions, the propulsion decrease, in case of moving fluid, contributes to experiment error [12]. Rotational Damping of STAR-CCM+ and experiment are compared in Fig. 7d. Notice that rotational CFD results can be improved by further simulations. In addition, as propulsion decrease and rope drag indicate a smaller hydrodynamic damping, therefore, we assume ideal nominal models by taking average of CFD and experiment results for every direction (Table 8). Nominals are used to demonstrate the nonlinear compensator in next section.

Table 8 Experimental results curve fitting (Benchmark: STAR-CCM+)

	Experiment & RMSE		STAR-CCM+ & RMSE	
Surge	$y = 21.4x^2 + 10.75x$	0.214	$y = 25.75x^2 + 0.2406x$	0.02294
Sway	$y = 61.39x^2 + 9.775x$	0.3817	$y = 48.39x^2 + 0.4512x$	0.0595
Heave(dive)	$y = 83.42x^2$	0.5924	$y = 82.44x^2$	0.1144
Yaw(left)	$y = 0.3513x^2 + 0.0321x$	0.119	$y = 0.1479x^2 + 0.001328x$	0.009881
Yaw(right)	$y = 0.3338x^2 + 0.1081x$	0.117	$y = 0.1479x^2 + 0.001328x$	0.009881
	Assumed nominal model			
Surge	$y = 25x^2 + 5.379x$			
Sway	$y = 57.48x^2 + 4.88x$			
Heave(dive)	$y = 80.37x^2$			
Yaw(left)	$y = 0.2496x^2 + 0.021x$			

5 Robust Control of CISCREA Yaw Heading

Without loss of generality, we demonstrate the robust controller in yaw direction. The rotational model is simplified as Eq. (18) (neglecting buoyancy and gravity). Definitions in yaw model are listed in Table 9.

$$(I_{YRB} + I_{YA})\ddot{x}_r + D_{YN}|\dot{x}|\dot{x} + D_{YL}\dot{x}_r = \tau_i \qquad (18)$$

As found in above sections, damping is a major nonlinearity in underwater vehicle models. We propose to compensate nonlinear behaviors by creating a linear behavior. The compensation error is assigned to be uncertainty.

Table 9 Rotational model notions of yaw direction

Parameter	Description	Value
I_{YRB}	Rigid-body inertia	$0.3578 \, \text{kg} \cdot \text{m}^2$
I_{YA}	Added mass inertia	$0.138 \, \text{kg} \cdot \text{m}^2$
D_{YN}	Nominal quadratic damping factors	Ideal 0.2496
D_{YL}	Nominal linear damping factors	Ideal 0.021
$\dot{x}_r$	Angular Velocity	0 to 4 rad/s
τ_i	Torque input	0 to 6 N $\cdot$ m
τ_{com}	Compensation Torque	0 to 6 N $\cdot$ m
$\dot{x}_{r0}$	Equilibrium velocity	0 to 4 rad/s
D_{YND}	CFD quadratic damping factors	0.1479
D_{YLD}	CFD linear damping factors	0.0013
D_{YLA}	Artificial linear factors	<Moto limit (select 1.2)

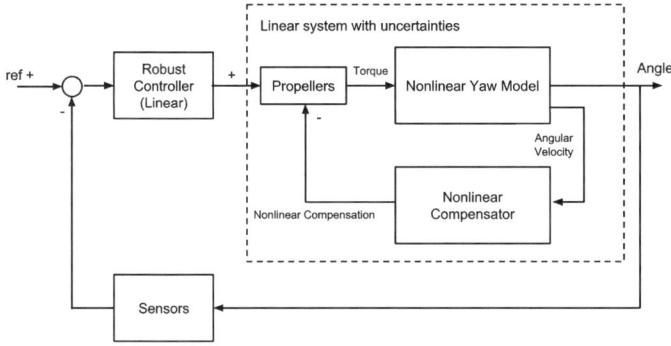

Fig. 8 Robust controller & Nonlinear compensator

Our approach uses nonlinear CFD yaw model as feedback to real world propellers, as shown in Fig. 8. The main idea is to compensate the original nonlinear behavior, and create a rough artificial linear damping behavior as robust control nominal. The nonlinear compensation is given in Eq. (19).

$$\tau_{com} = (D_{YLA} - D_{YLD} - D_{YND}|\dot{x}_r|)\dot{x}_r \tag{19}$$

D_{YLA} is the artificial linear factor given in Table 9. D_{YND} and D_{YLD} are CFD damping estimations. The linear model result of compensation is given in equation (20).

$$(I_{YRB} + I_{YA})\ddot{x}_r + (D_{YLA} + (D_{YN}|\dot{x}_r| - D_{YND}|\dot{x}_r| + D_{YL} - D_{YLD}))\dot{x}_r = \tau_i \tag{20}$$

The term $\delta = D_{YN}|\dot{x}_r| - D_{YND}|\dot{x}_r| + D_{YL} - D_{YLD}$ is calculated as an uncertainty added on D_{YLA}. Generally, this δ is small comparing to D_{YLA}.

If we calculate δ using that,

$$\dot{x}_r \in [-4, 4] \text{ rad/s}; D_{YLA} = 1.8; D_{YN} = 0.2496;$$
$$D_{YL} = 0.021; D_{YND} = 0.1479; D_{YLD} = 0.0013;$$

we can then consider that D_{YLA} has a dynamic uncertainty of 23.7%, listed in Table 10. At the end, the proposed model, Eq. (21) is a first order linear system.

$$(I_{YRB} + I_{YA})\ddot{x}_r + (D_{YLA} + \delta)\dot{x}_r = \tau_i; \delta \in [-0.4265, 0.4265] \tag{21}$$

Table 10 Linear damping uncertainty margin

Methods	Nominal linear factor	Uncertainty margin
Compensate	D_{YLA}:1.8 (for example)	D_{YLA} : [1.3735, 2.2265], 23.7%

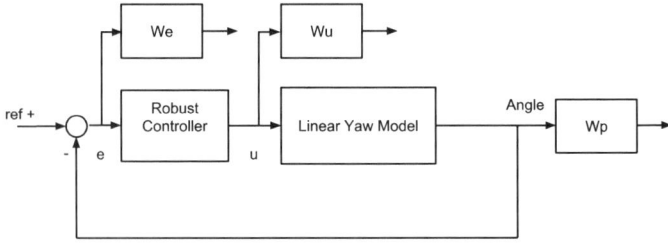

Fig. 9 Weighting functions for robust synthesis

The robust controller is based on H_∞ control design [26], see Fig. 9. The weighting functions are given as [13]:

$$W_p(s) = Gp \frac{s^2 + p_{n1}s + p_{n2}}{s^2 + p_{d1}s + p_{d2}} \tag{22}$$

$$W_e(s) = Ge \frac{s + e_{n1}}{s + e_{d1}} \tag{23}$$

$$W_u(s) = Gu \tag{24}$$

Controller synthesis configurations are listed in Table 11. Step responses of PID controller, robust controller with and without nonlinear compensation are compared in Fig. 10. We should highlight that our H_∞ controller handle the nonlinearity with faster response.

Table 11 Robust synthesis configurations

Parameters	Value
Inertia variation	30 %
Damping variation	50 %
Synthesis algorithm	LMI
p_{n1}	1.8
p_{n2}	10
p_{d1}	8
p_{d2}	0.01
e_{n1}	0.92
e_{d1}	0.0046
Gp	0.95
Ge	0.5
Gu	0.01

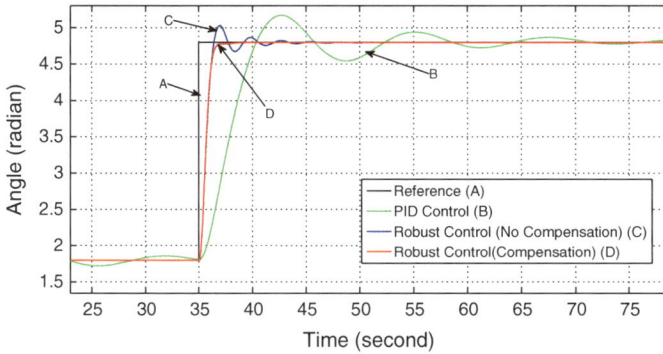

Fig. 10 Yaw controlled response

6 Conclusion

In this manuscript, an AUV modeling approach is proposed to avoid the deployment of expensive devices. The quantitative model is built for CISCREA, and validated by realistic experiments. We estimated numerically two important hydrodynamic parameters: the added mass (Predicting by WAMIT™ and MCC) and the damping effects (Predicting by ANSYS-CFX™, STAR-CCM+™ and experiments). Our experiment results showed that the quadratic damping is the dominant component of all damping. With our numerical model, we propose nonlinear compensator to shrink the uncertainty margin for linear-based robust control designs. Note that with even our unimproved rotational CFD results, we can guarantee an uncertainty margin less than 23.7 % (linear damping factor). In the end, we simulated the model based yaw robust controller in Matlab. Our controller shows no oscillation in step response, and it is faster than PID controller. Our results have been validated in real test, this is beyond the scope of this manuscript, and it will be published in future works.

Acknowledgments The authors would like to express their great appreciation to the China Scholarship Council for their financial supports and to Prof. J. M. Laurens and F. Le Bars for their technical supports to complete this work successfully.

References

1. Caccia, M., Indiveri, G., & Veruggio, G. (2000). Modeling and identification of open-frame variable configuration unmanned underwater vehicles. *IEEE Journal of Oceanic Engineering*, 25(2), 227–240.
2. Clement, B. (2012). Interval analysis and convex optimization to solve a robust constraint feasibility problem. *European Journal of Automated Systemes*, 46(4–5), 381–395.
3. Comolet, R., & Bonnin, J. (1979). *Experimental fluid Mechanics. Tome I*. Paris: Masson.
4. Doyle, J. C. (1978). Guaranteed margins for lqg regulators. *IEEE Transactions on Automatic Control*, 23, 756–757.

5. Eng, Y. H., Lau, W. S., Low, E., & Seet, G. G. L. (2008). Identification of the hydrodynamics coefficients of an underwater vehicle using free decay pendulum motion. In *International MultiConference of Engineers and Computer Scientists* (vol. 2, pp. 423–430). HongKong, March 2008.
6. Eng, Y. H., Lau, M. W., & Chin, C. S. (2013). Added mass computation for control of an openframe remotely-operated vehicle: Application using wamit and matlab. *Journal of Marine Science and Technology, 22*(2), 1–14.
7. euRathlon: Outdoor robotics challenge for land, sea and air. (2014). Retrieved September 15, 2014, from http://www.eurathlon.eu/site/.
8. Feng, Z., & Allen, R. (2004). Reduced order h1 control of an autonomous underwater vehicle. *Control Engineering Practice, 12*(12), 1511–1520.
9. Ferreira, B. M., Matos, A. C., & Cruz, N. A. (2012). Modeling and control of trimares auv. In *12th International Conference on Autonomous Robot Systems and Competitions* (pp. 57–62). Portugal, April 2012.
10. Ferreira, B., Pinto, M., Matos, A., & Cruz, N. (2009). Hydrodynamic modeling and motion limits of auv mares. In *35th Annual Conference of IEEE on Industrial Electronics* (pp. 2241–2246). Porto, November 2009.
11. Fossen, T. I. (1994). *Guidance and control of ocean vehicles*. New York: Wiley.
12. Fossen, T. I. (2002). Marine control systems: Guidance, navigation and control of ships, rigs and underwater vehicles. *Marine Cybernetics Trondheim.*
13. Gu, D. W., Petkov, P. H., & Konstantinov, M. M. (2005). *Robust control design with MATLAB.* Springer.
14. Jaulin, L., & Bars, F. L. (2012). An interval approach for stability analysis: Application to sailboat robotics. *IEEE Transaction on Robotics, 27*(5), 282–287.
15. Maalouf, D., Tamanaja, I., Campos, E., Chemori, A., Creuze, V., Torres, J., & Rogelio, L., et al. (2013). From pd to nonlinear adaptive depth-control of a tethered autonomous underwater vehicle. In *5th Symposium on System Structure and Control*. Grenoble, France, February 2013.
16. MCC: Marine craft characteristics. (2014). Retrieved February 01, 2014, from https://moodle.ensta-bretagne.fr/course/view.php?id=18.
17. MSS: Marine systems simulator. (2010). Retrieved Feruary 01, 2014, from http://www.marinecontrol.org.
18. Perrault, D., Bose, N., O'Young, S., & Williams, C. D. (2003). Sensitivity of auv added mass coefficients to variations in hull and control plane geometry. *Ocean Engineering, 30*(5), 645–671.
19. Rentschler, M. E., Hover, F. S., & Chryssostomidis, C. (2003). Modeling and control of an odyssey iii auv through system identification tests. In *Unmanned Untethered Submersible Technology Conference*. Durham, USA, August 2003.
20. Roche, E., Sename, O., & Simon, D. (2010). Lft/h1 varying sampling control for autonomous underwater vehicles. In *4th IFAC Symposium on System, Structure and Control* (pp. 17–24). Delle Marche, Italy, September 2010.
21. Ross, A., Fossen, T. I., & Johansen, T. A. (2004). Identification of underwater vehicle hydrodynamic coefficients using free decay tests. In *IFAC Conference on Control Applications in Marine Systems*. Ancona, Italy, July 2004.
22. SAUC-E: Student autonomous underwater challenge-europe. (2014). Retrieved September 19, 2014, from http://sauc-europe.org.
23. SNAME [1950]: (1950). Nomenclature for treating the motion of a submerged body through a fluid. In *The Society of Naval Architects and Marine Engineers, Technical and Reserach Bulletin* (pp. 1–15), April 1950
24. Yamamoto, I. (2001). Robust and non-linear control of marine system. *International Journal of Robust and Nonlinear Control, 11*(13), 1285–1341.
25. Yang, R., Clement, B., Mansour, A., Li, H. J., Li, M., Wu, N. L. (2014). Modeling of a complex-shaped underwater vehicle. In *2014 IEEE International Conference on Autonomous Robot Systems and Competitions (ICARSC2014)*. Espinho, Portugal, May 2014.
26. Zhou, K. M., & Doyle, J. C. (1998). *Essentials of robust control*. Prentice hall.

SAMDIS: A New SAS Imaging System for AUV

Myriam Chabah, Nicolas Burlet, Jean-Philippe Malkasse,
Guy Le Bihan and Bruno Quellec

Abstract Synthetic Aperture Sonars (SAS) image sea floor at high resolution, independently of range. They equip towed bodies as well as Autonomous Underwater Vehicles (AUV), surveying seabed in total autonomy. Thales Underwater Systems developed a new sonar, named SAMDIS (Synthetic Aperture Mine Detection and Imaging System), a compact system which can be easily mounted on AUV and delivers high-resolution underwater SAS imagery. In order to change the aspect angle of a scene without wasting time in revisiting the place, the SAMDIS sonar processes SAS imagery under different view angles simultaneously. The diversity of aspect angles associated with high resolution SAS image is mandatory for Mine Counter Measure applications notably for classification. Furthermore, the SAMDIS sonar is equipped with an interferometric antenna, similar to the imaging antenna. It provides SAS bathymetry maps at high-resolution and enhances classification probability in non-flat sea bottom configuration. In this paper, we describe the SAMDIS complete solution of high resolution SAS multi-aspect and SAS interferometry, which was tested during first sea trials.

M. Chabah (✉) · N. Burlet · J.-P. Malkasse · G. Le Bihan · B. Quellec
Thales Underwater Systems, Route de Sainte Anne du Portzic, CS 43814,
29238 Brest Cedex 3, France
e-mail: myriam.chabah@fr.thalesgroup.com

N. Burlet
e-mail: nicolas.burlet@fr.thalesgroup.com

J.-P. Malkasse
e-mail: jean-philippe.Malkasse@fr.thalesgroup.com

G. Le Bihan
e-mail: guy.lebihan@fr.thalesgroup.com

B. Quellec
e-mail: bruno.quellec@fr.thalesgroup.com

© Springer International Publishing Switzerland 2016
B. Zerr et al. (eds.), *Quantitative Monitoring of the Underwater Environment*,
Ocean Engineering & Oceanography 6, DOI 10.1007/978-3-319-32107-3_10

1   Introduction

Synthetic Aperture Sonar (SAS) imaging is today a mature technology, used in operational systems equipped with side looking sonar. SAS processing creates a synthetic antenna, longer than the real antenna, by adding pings coherently along the sonar displacement, achieving a better resolution than the physical one [1].

Classification, particularly within the framework of mine hunting, benefits from this resolution gain [2] and new products have emerged in order to extend the use of SAS processing. Hence, instead of the traditional broadside SAS mode, a squint spot SAS mode was implanted in a Mine Counter measure Vessel equipped with a front looking sonar [3].

However, in order to further improve classification, there is a need to take into account more information, at high resolution. For this purpose, imaging one object from different points of views [4, 5], or computing high resolution bathymetric information [6] are of deep interest.

Thales Underwater Systems has developed a new sonar, SAMDIS, gathering these new functionalities, multi-aspect SAS and SAS interferometry. First sea trials have allowed analyzing their performance on real data.

In the first part, SAMDIS system and sea trials are presented, multi-aspect SAS and SAS interferometry principles are described in the second and third part. In each part, results are given and real data images illustrate the way these new functions can help with the classification step.

2   SAMDIS System

SAMDIS is a Synthetic Aperture Mine Detection and Imaging System developed by Thales Underwater Systems. Its wideband interferometric Synthetic Aperture Sonar, consists of two along-track receiver arrays, on each side, providing a vertical interferometric baseline. Multipath effect in shallow water is countered by narrow vertical beam patterns.

SAMDIS provides a large panel of real-time, high resolution imaging functions (Fig. 1). Combination of real-time high resolution multi-aspect and SAS interferometry waterfalls results in enhanced performance in scene interpretation and benefits to the detection, classification and localization of potential mines. Post-mission data processing and tactical management suite presents all these information, including Automatic Target Recognition contacts, in an efficient way to the operator (Fig. 2).

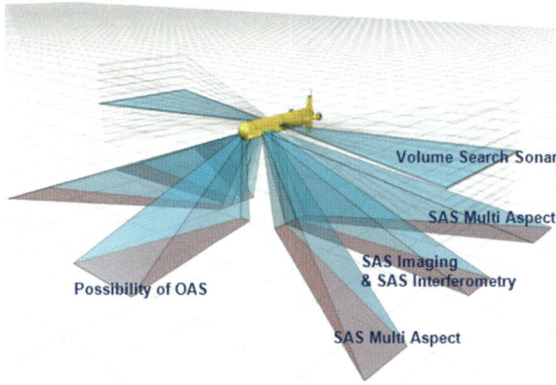

Fig. 1 SAMDIS imaging capabilities

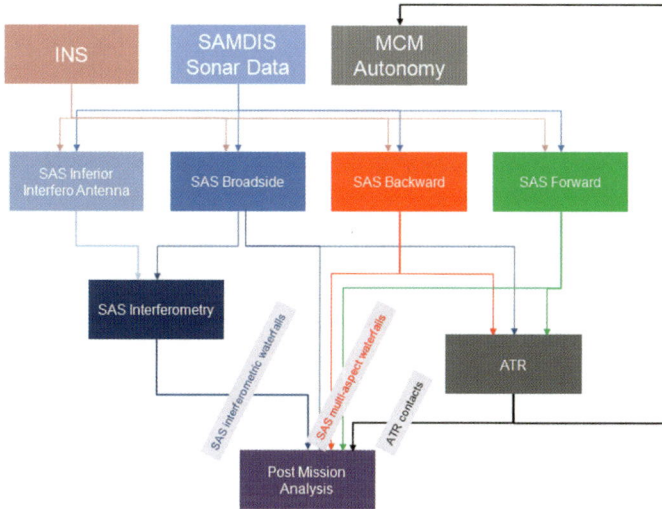

Fig. 2 Overview diagram of SAMDIS MCM solution

SAMDIS payload is available in a series of different lengths for adaptation to platforms and mission. It can be easily integrated on towed body or on AUV (Fig. 3).

During summer 2014, sea trials have been performed in Douarnenez Bay, France. SAMDIS was mounted on an AUV. Several tracks were realized over prelaid targets, allowing testing the imaging ability of the new payload. In this area, water depth was fluctuating around 30 m.

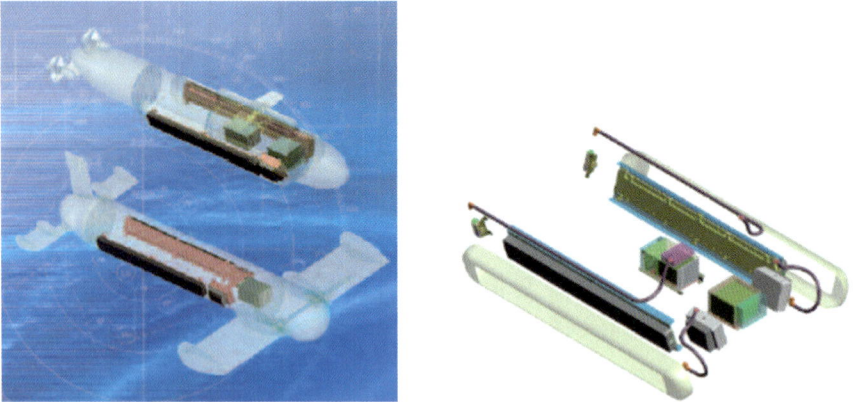

Fig. 3 SAMDIS is easily integrated on AUV or TSAS

3 Multi-aspect Imaging

3.1 Principle

Conventional MCM vessels are used to turn around a suspicious object in order to collect independent information and classify. Multi-aspect SAS imaging offers this capacity at high resolution and without wasting time in revisiting the area: SAMDIS enables to process three different views simultaneously (one standard broadside image, and two squinted images, Fig. 4). Indeed, three SAS waterfalls are computed in parallel, with identical high resolution given by hybrid Ping to Ping Cross Correlation and SAS processing [7, 8].

The operational interest is high as a single track gives three aspects of every object.

Fig. 4 Multi-aspect SAS geometry

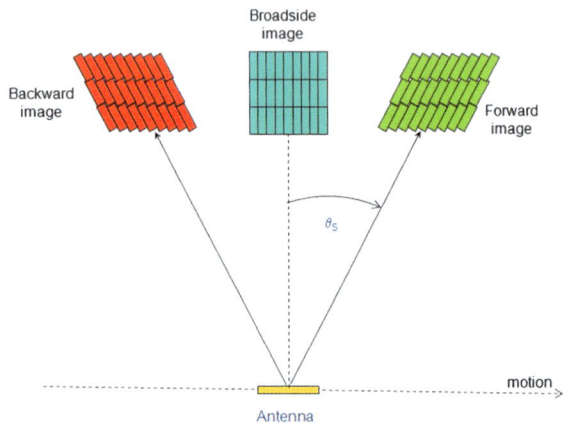

Fig. 5 Rock hiding a mine.
The mine can only be
detected in squinted images

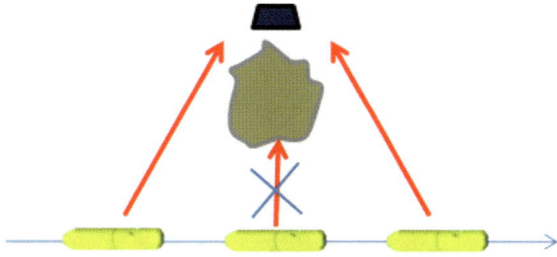

Fig. 5 Rock hiding a mine.
The mine can only be
detected in squinted images

Let's consider the case of a mine which is hidden behind a rock (Fig. 5). It would not even be detected in a broadside-only process. In a more general way, multi-aspect SAS significantly increases the confidence for any classification analysis. In mine hunting, the key point is to decrease the classification ambiguity factor [9], that is to say, the amount of non-mine contacts with the same statistical characteristics than those of the searched mines, which are responsible for the false-true classification probability. The first way to lower the classification ambiguity is to improve the resolution, and the second one is to increase the number of independent views of a same contact. Multi-aspect SAS imaging allows doing both: it reaches a high clearance rate, that is to say a high classification probability, with low false-true classification probability, and this is done in a limited time matching with operational time constraints.

It is all the more significant in conditions where a variation of track orientation is difficult or impossible. This is the case when water current imposes the track direction or when "channel like" areas have to be investigated. More generally, for task optimization reasons, multiple U-turns are not recommended as they decrease the efficiency of the mission [10].

3.2 Results

Three extracts of multi-aspect waterfalls are shown in Fig. 6. Backward, broadside and forward SAS images represent a same area.

Figure 7 shows the backward, broadside and forward images of an object. The shadowvaries a lot from one aspect angle to another, as it is the case for non-symmetricalobjects. However, for each aspect, the shadow fits the expected shadow very well,increasing the classification confidence.

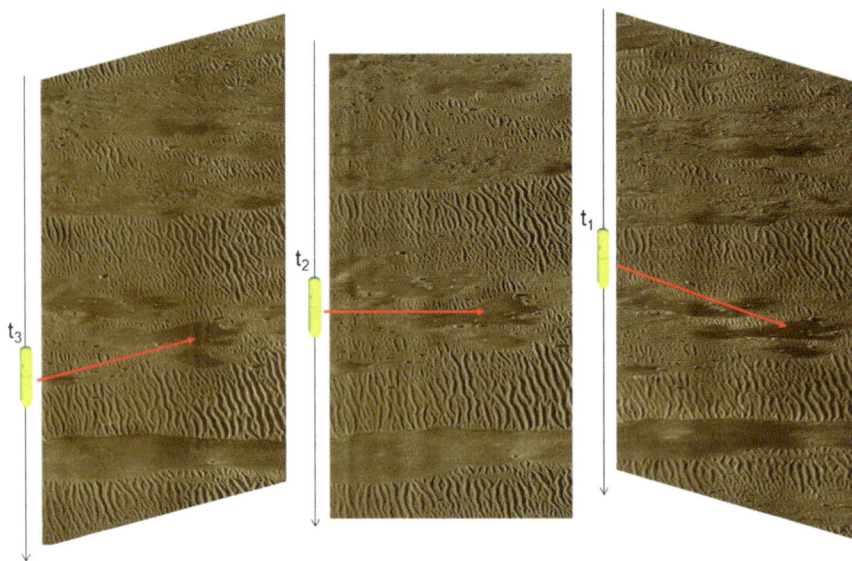

Fig. 6 Multiple-aspect waterfalls. On the *left*, backward SAS image, in the *middle*, broadside SAS image and on the *right*, forward SAS image

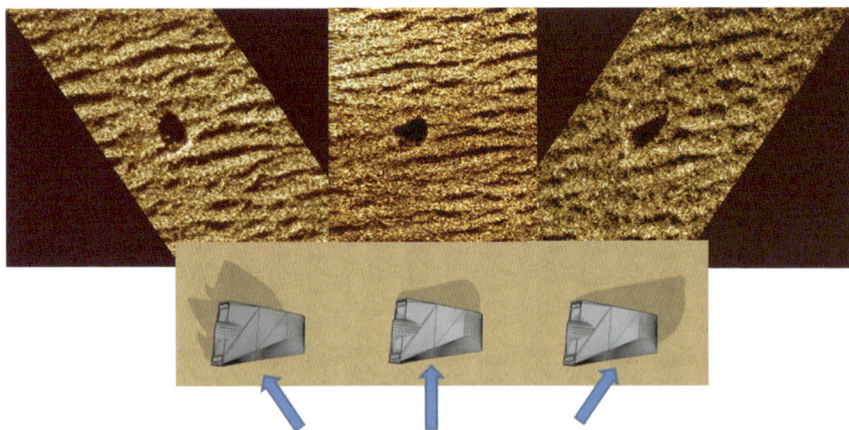

Fig. 7 Multiple-aspect SAS images of an asymmetrical object

4 SAS Interferometry

4.1 Principle

Interferometry is based on the estimation of the difference of travel paths of an incident wave, received at least at two vertically spaced receivers (Fig. 8). This path difference or, equally, the delay $\Delta\tau$, difference of times of arrival, is proportional to the cosine of the direction of arrival, α, of the insonified zone (1).

$$\cos(\alpha) = \frac{(r_1 - r_2)}{L_B} = \frac{C}{L_B}\Delta\tau \qquad (1)$$

Geometric considerations enable to retrieve the difference of height between the center of the interferometric base and the insonified target (2), taking into account the angle, ψ, between the interferometric base plane and the vertical plane.

$$\Delta H = r\cos(\alpha - \psi) \qquad (2)$$

Typically, interferometry is computed with a base consisting of two physical antennas. At the condition that the SAS imaging process is phase preserving, interferometry can be computed with a base consisting of two synthetic antennas [11, 12]: interferometric SAS capability was demonstrated at sea in 2002 with Imbat, a towed interferometric SAS sonar developed by Thales Underwater Systems [13, 14].

In order to estimate the delay $\Delta\tau$, special attention must be paid to the co-registration of the SAS data received by the two antennas. Then, phase estimation is processed, at high resolution, followed by a 2D unwrapping technique.

The sampling of the interferometric map is customizable, depending on the vertical resolution that is needed [8].

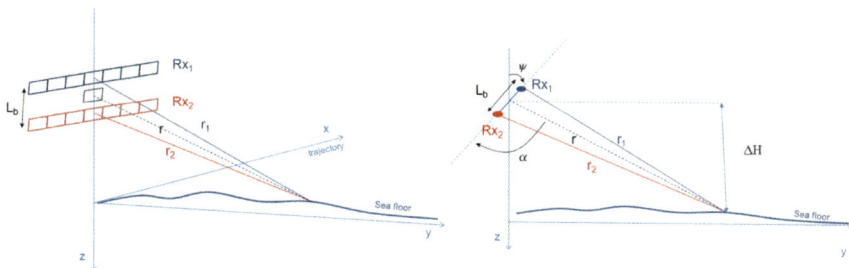

Fig. 8 Interferometry principle

4.2 Results

A first example of SAMDIS interferometry ability is given in Fig. 9 and the corresponding SAS image can be seen in Fig. 10, where ripples of 10 cm are observed, at a range of approximately 40 m. At this short range, with a sonar height of 15 m, shadowing has reduced effect and bathymetry can be estimated even in the trough of the ripples.

The second example concerns an inclined plane, which has been insonified (Fig. 11). The SAS image is given in Fig. 12 and the corresponding bathymetry in Fig. 14. The dimensions of the inclined plan estimated on the interferometric SAS

Fig. 9 SAMDIS interferometric SAS image of ripples in Douarnenez Bay

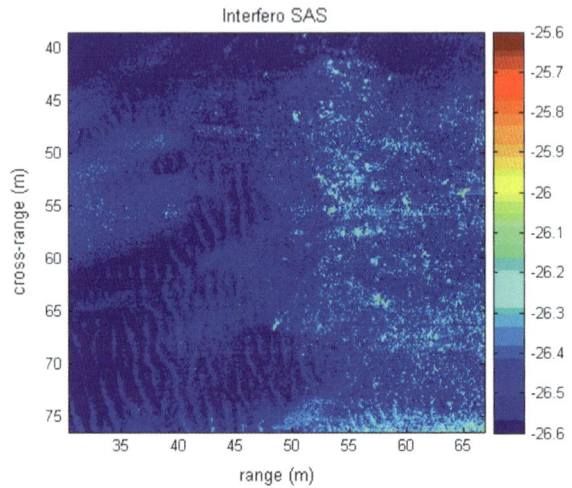

Fig. 10 SAMDIS SAS image of ripples in Douarnenez Bay

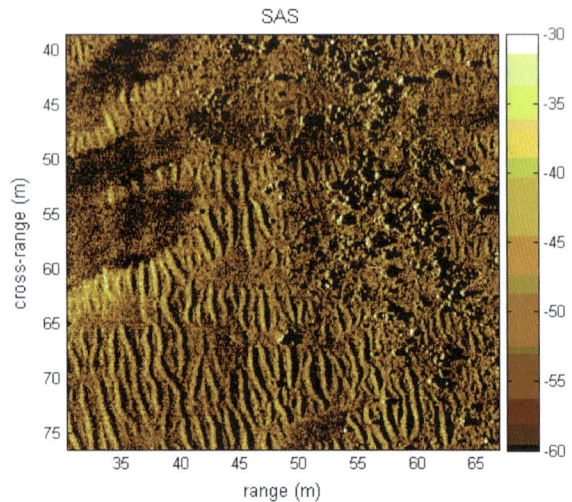

Fig. 11 Inclined plane

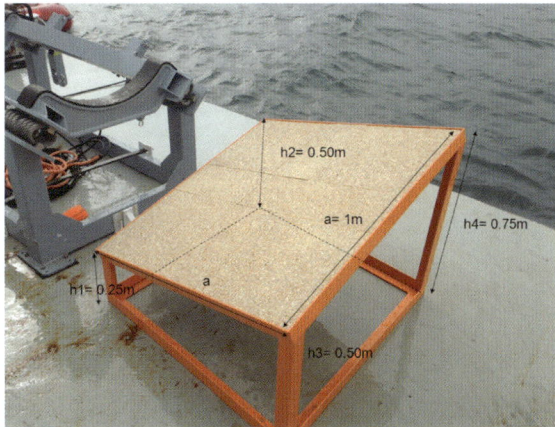

Fig. 12 Inclined plan SAS image

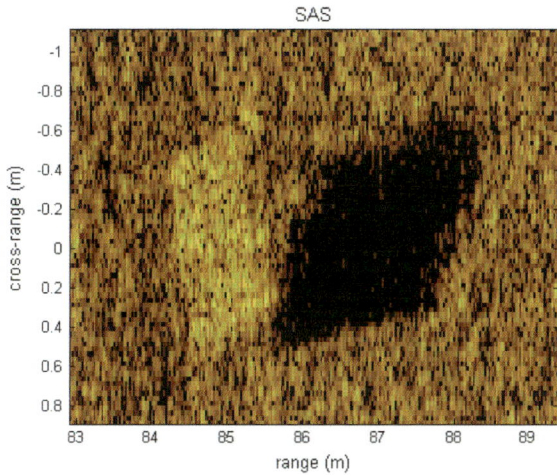

Fig. 13 SAS bathymetric image of the inclined plan. No noise reduction filtering has been applied in order to observe the resolution performance. The color bar ranges from −32.8 to −32.2 meters, and shows that the estimated difference of elevations of the inclined plan is close to the expected 50 cm

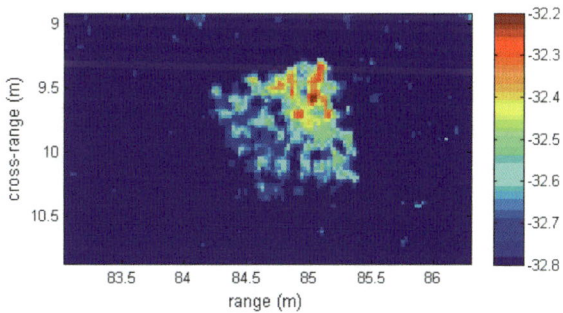

Fig. 14 Interferometric SAS
image of the inclined plan.
The non-significant values of
the shadow area have been
removed

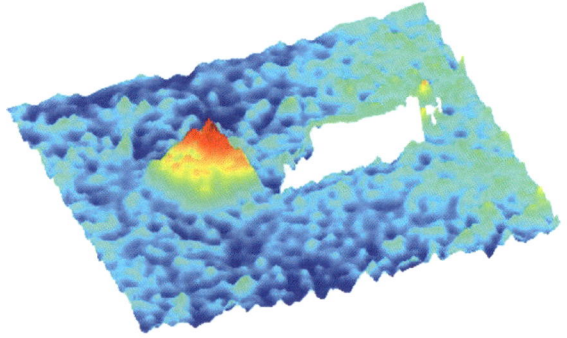

Fig. 15 SAS texture and
interferometric SAS elevation
in the vicinity of a moored
mine. Both float and sinker
echoes can be observed, as
well as the float shadow. The
dimension of the zoom along
the x axis is 2 meters long,
and 20 m along the y axis.
The elevation of the pic is
88 cm

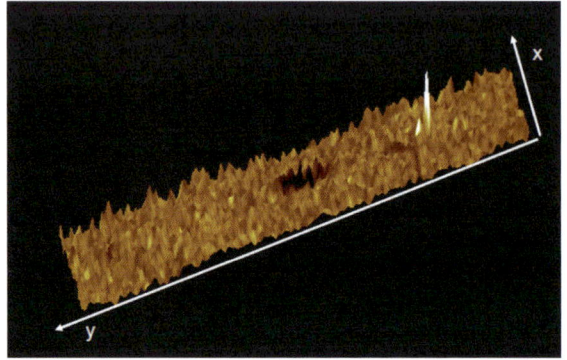

image, computed through the bathymetry of the four corners are close to reality, with an error inferior to 5 cm (Fig. 13).

High resolution bathymetry enhances classification confidence. Figure 15 shows the bathymetry on a zone surrounding a moored mine with a tether of 70 cm.

In this example, the relative position between the echo and the shadow on the SAS image enables to classify the object as a moored-mine and to give an estimation of the tether length, L. The interferometric SAS image confirms these results, and, giving the position of the scattering on the sphere, the estimated tether length is computed with an error inferior to 5 cm (Fig. 16).

In the last example, illustration is given below that classification would have failed without the bathymetry information. In the following example, a mine is positioned on the ascending slope of a bump. Without bathymetric information and assuming a flat sea bottom, the estimated height of the target on the basis of the length of the shadow would be around 30 cm. Taking into account the 5 % slope, given by the bathymetry map, the estimated height would be 41 cm, which is close to the value given by the bathymetry map of the summit of the mine, and close to the real height of 38 cm. Higher object or steeper slope would enhance the error due to the flat sea bottom assumption (Fig. 17).

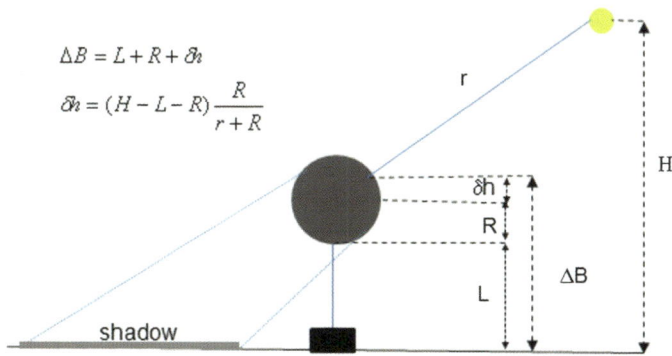

$$\Delta B = L + R + \delta h$$

$$\delta h = (H - L - R)\frac{R}{r + R}$$

Fig. 16 Geometric analysis of the moored mine scene

Fig. 17 SAS image texture combined with bathymetric elevation

5   Conclusion

Thales Underwater Systems has developed a new compact sonar system, which provides high resolution SAS multi-aspect real-time waterfalls and SAS interferometry waterfalls.

This new system was tested during first sea trials. It is operational and offers a complete solution for mine hunting missions.

High resolution interferometry gives a measure of an object dimensions and bathymetry information enables to take into account the distortion of shadows, according to the relief of the sea bottom.

High resolution multi-aspect SAS computes independent images of a same object during a single track and enables to combine both time-tested mine hunting strategy of conventional MCM vessels and efficiency of high resolution SAS.

In that way, it achieves a high classification probability in a limited clearance time, making multi-aspect SAS mandatory for efficient SAS mine hunting.

Work continues on different subjects benefitting from the new SAMDIS sensors:

- Information presentation to operators in an ergonomic exploitation tool and information combination into operator aid functions.
- Enhancement of resulting autonomy for AUV vehicles and mission planning.
- Change detection in surveillance situation

Acknowledgments The authors wish to acknowledge DGA/UM-NAV, French Marine Nationale and DGA/GESMA Brest Center for sea trials support.

References

1. Billon, D., & Fohanno, F. (1998). Theoretical performance and experimental results for synthetic aperture sonar self-calibration. In *Proceedings of Oceans'98* (pp. 965–970). Nice, France.
2. Florin, F., Fohanno, F., Quidu, I., & Malkasse, J. P. (2004). Synthetic aperture and 3D imaging for mine hunting sonar. In *UDT Europe 2004*.
3. Burlet, N., Chabah, M., & Poirier, E. (2010). New synthetic aperture sonar classification mode for mine countermeasure vessels. In *Proceedings—Institute Of Acoustics-Cd-Rom Edition-;* 32; P4 (International Conference SAS and SAR, 13–14 September 2010, Italy).
4. Leblond, I., & Quidu, I. (2010). Mise en oeuvre de stratégies prédictives sur les vues à acquérir pour la classification multi-vue d'objets immergés à partir d'images SAS. In *RFIA 2010*.
5. Fernandez, J. E., & Christoff, J. T. (2000). Multi aspect synthetic aperture sonar. In *Proceedings of Oceans 2000.* Providence, RI USA, September 2000.
6. Hansen, R. E., & Saebø, T. O., Gade, K., & Chapman, S. (2003). Signal processing for AUV based interferometric synthetic aperture sonar. In *Proceedings of Oceans 2003.* MTS/IEEE (pp. 2438–2444). San Diego, CA, USA, September 2003.
7. Billon, D., & Fohanno, F. (1998). Theoretical performance and experimental results for synthetic aperture sonar self-calibration. In *Oceans'98* (pp. 965–970), September 1998.
8. Billon, D. (2000) Interferometric synthetic aperture sonar, design and performance issues. In *Proceedings of th 5th European Conference on Underwater Acoustics, ECUA 2000.*
9. Florin, F., Quidu, I., & Malkasse, J. P. (2005). From UUV to AUV: the SCM challenge. In *Undersea Defence Technology (UDT) Europe 2005*, Amsterdam, June 2005.
10. Arnold-Bos, A., Bacor, I., Brunet, J. P., & Lecouvez, M. Theatre-wide automatic mission scheduling for a mine countermeasure force. In *Undersea Defence Technology (UDT) Europe 2012*, Alicante, May 2012.
11. Barclay, P. J. (2006). Interferometric Synthetic Aperture Sonar Design and Performance, thesis presented for the degree of Doctor of Philosophy in Electrical and Computer Engineering at the University of Canterbury, Christchurch, New Zealand, August 2006.
12. Saebø, T. O. (2010). Seafloor depth estimation by means of interferometric Synthetic Aperture Sonar (University of Tromsø, september 2010). *Dissertation for the degree of Philosophiae Doctor.*
13. Bréchet, D., Billon, D., & Fohanno, F. (2002). Results from imbat, a deep sea mapping sonar system. In *6th workshop on underwater acoustics* (pp. B1–B15), Brest, June 2002.
14. Billon, D. (2005). About accuracy of the elevation angle measurement in interferometric synthetic aperture sonar. In *Oceans 2005 Europe* (vol. 1, pp. 650–654), June 2005.

Obstacle Avoidance for an Autonomous Marine Robot—A Vector Field Approach

Silke Schmitt, Fabrice Le Bars, Luc Jaulin and Thomas Latzel

Abstract A marine robot, especially a sailing boat robot, is exposed to a dynamic environment. This paper presents a simple and efficient obstacle avoidance control algorithm. The presented control method uses vector fields to regulate the marine robot.

1 Introduction

There is a growing interest in autonomous marine robots, for example they can be used for measurements on the sea or harbour monitoring [1, 2]. An autonomous marine robot has several advantages over oceanographic boats (with a crew) and buoys; these are explained in paper [1]. For missions of autonomous marine robots it is important to have an efficient and reliable control algorithm, especially when thinking of long term missions [3]. The detection of moving obstacles is possible using a number of technologies such as radar, camera and an automatic identification system (AIS). The algorithm should not just consider the desired path of the robot but should also detect other obstacles and avoid them. Moreover a general approach is desired, which cannot just be applied on one special type of marine robot but can be adapted to any kind of marine robot.

A control method that uses vector fields fulfills these requirements. The construction and application of a vector field to regulate a marine robot is presented in this paper. This approach of a potential field method is not very common on marine robots

S. Schmitt (✉) · T. Latzel
Universitaet der Bundeswehr, Werner-Heisenberg-Weg 39, 85579 Neubiberg, Germany
e-mail: s1lke.schmitt@t-online.de

T. Latzel
e-mail: thomas.latzel@unibw.de

F. Le Bars · L. Jaulin
ENSTA Bretagne, 2 rue Francois Verny, 29806 Brest Cedex 9, France
e-mail: Fabrice.LE_BARS@ensta-bretagne.fr

L. Jaulin
e-mail: Luc.JAULIN@ensta-bretagne.fr

© Springer International Publishing Switzerland 2016
B. Zerr et al. (eds.), *Quantitative Monitoring of the Underwater Environment*,
Ocean Engineering & Oceanography 6, DOI 10.1007/978-3-319-32107-3_11

yet but is very appealing due to its simplicity. Assuming that a marine robot usually has a very limited processing power and sensing, it is important to consider the complexity of the control algorithm. The obstacle avoidance control algorithm was simulated with sailboat robots. The functionality of the obstacle avoidance control algorithm has been validated during WRSC 2013 where the algorithm was applied on the autonomous motorboat of the Team ENSTA Bretagne—Ifremer.

2 Basis of the Control Method: Vector Fields

In a vector field [4] there is an assignment of a vector to each point. In the case of this application a point (x, y) is mapped to a vector $\mathbf{f}(x, y)$. This is a function $\mathbf{f}$ of the form $\mathbf{f} \colon \mathbb{R}^2 \to \mathbb{R}^2$ [5]. For the control method these vectors are considered as forces. The boat robot (represented as a point) moves in this vector field along the vector gradient, so the vector field describes the behaviour of the marine robot.

In order to be able to follow different trajectories, there has to be the possibility of constructing a complex vector field. This is why the vector field is constructed as a binary tree. On the basis of simple atomic vector fields (e.g. a vector field where all vectors point towards one line) which can be combined, it is possible to get a more complex vector field. Merging vector fields is always done in building a binary tree, where the top node (thus the entire tree) is the desired complex vector field.

2.1 Atomic Vector Fields

Two types of atomic vector fields can be distinguished. The first is an attractive field, where the vectors of the field point towards a desired trajectory. The second type is a repulsive vector field, in which the vectors point away from a particular area.

The vectors are normalized to the same limit values in their magnitude (when not going to infinity which is pointed out when explaining the specific vector field); like this they will have the same impact when they are combined by an operation.

As attractive vector field have been implemented:

- a field for an attraction to a circle (Fig. 1)
- a circular vector field (Fig. 2)
- an attraction towards a line (Fig. 3)
- a vector field with vectors into one direction (Fig. 4)
- a vector field for an attractive point in a long distance version (Fig. 5)
- a vector field for an attractive point in a short distance version (Fig. 6)

These fields are presented in the following with an explanation, an equation and a figure.

Jaulin [6] provides an equation for a circle following (a vector field where every vector points towards the desired path of a circle). In fact this is a composition of

Fig. 1 Attractive circle

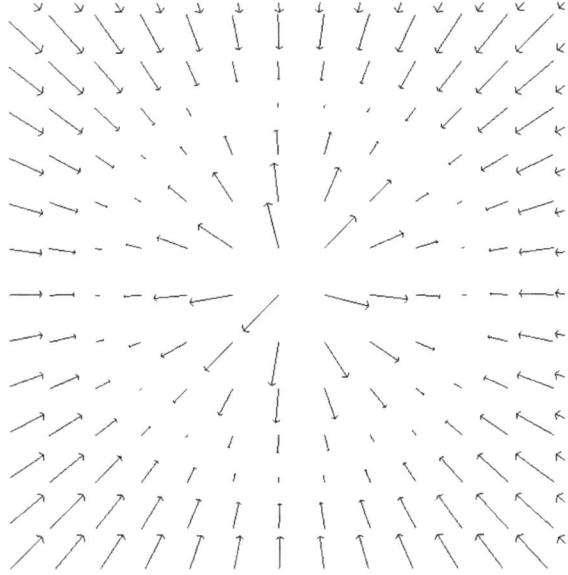

Fig. 2 Circular vector field

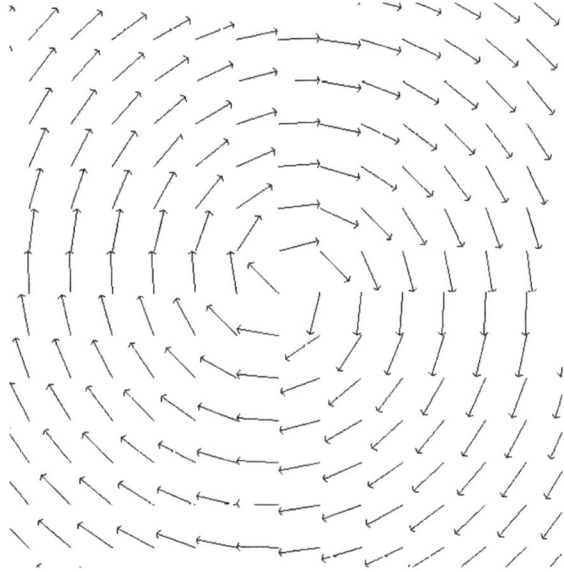

Fig. 3 Attractive line

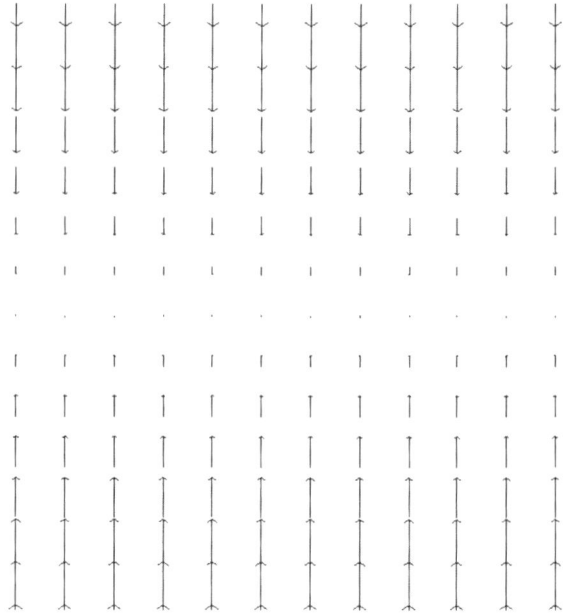

Fig. 4 GoX vector field

Fig. 5 Attractive
point—long range

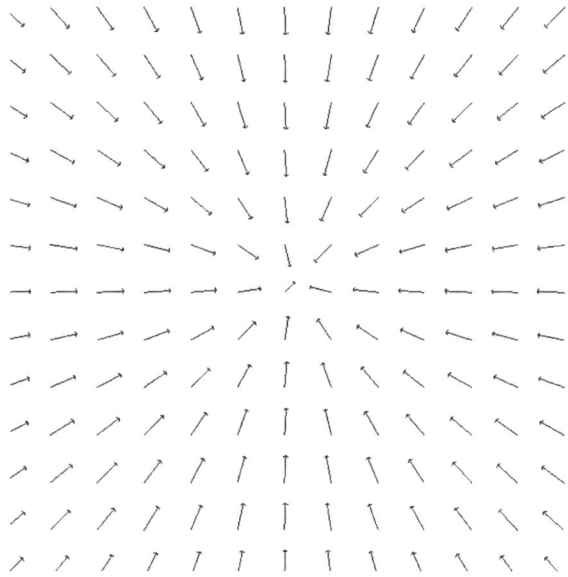

Fig. 6 Attractive
point—short range

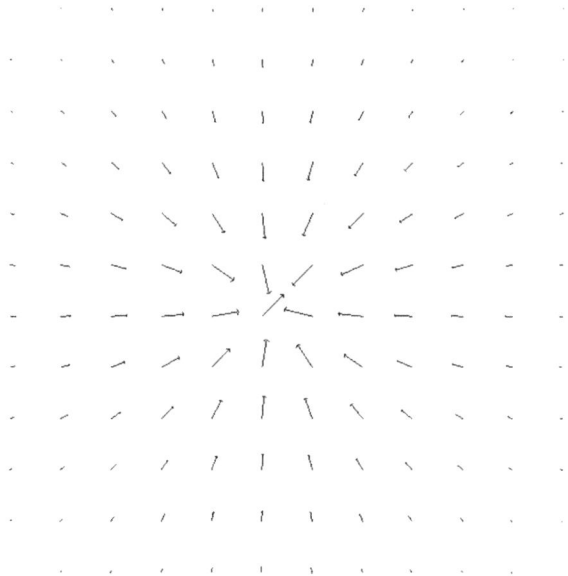

two basic vector fields, an attraction towards a circle with a radius r_0 and a parameter $\rho \in [0, 1]$ that defines how fast the final state (circle) is approached and a circular vector field. To have a set of combinable atomic vector fields, the two vector fields are implemented seperately. Equation (1) is the function for an attraction towards a circle and Eq. (2) is a circular vector field.

$$\mathbf{f}(x, y) = \begin{pmatrix} x \arctan(\rho(r_0 - \sqrt{x^2 + y^2})) \\ y \arctan(\rho(r_0 - \sqrt{x^2 + y^2})) \end{pmatrix} \tag{1}$$

$$\mathbf{f}(x, y) = \begin{pmatrix} y \\ -x \end{pmatrix} \tag{2}$$

A line following is a similar combination of two atomic vector fields. Jaulin and Bars [7] presents a function to regulate the boat along a line from point A to B (infinite line). Here the two seperate parts are implemented independently. Equation (3) shows the attraction towards a line on the x-axis and Eq. (4) is a constant vector field to the east.

$$\mathbf{f}(x, y) = \begin{pmatrix} 0.0 \\ \frac{1}{y} \cdot (10 * y * \arctan(-\frac{y}{50})) \end{pmatrix} \tag{3}$$

$$\mathbf{f}(x, y) = \begin{pmatrix} 10.0 \\ 0.0 \end{pmatrix} \tag{4}$$

Without loss of generality the direction of these two fields can be chosen to be fixed; the position and orientation may be changed by methods (rotation and shift) which can be applied to these atomic vector fields. For a line following task these two vector fields are added; the vector orientation is parallel to the line, when the boat position is on the line. Otherwise the angle is inclined towards the line; the greater the distance to the line, the higher the inclination. There is a maximum inclination of 45° (far away from the line where the vectors of both atomic fields have the same magnitude), so that the boat still keeps going ahead in the direction of the line.

A vector field towards one point, a long-range attractive point, is useful to have one goal position which is approached no matter where the boat is located. The function $f(x, y)$ to calculate the vector magnitude has to fulfill the requirements of having a constant limit value for $x \to \infty$ and $y \to \infty$ as well as going through the point $f(0, 0) = 0$. The vector magnitude is nearly constant. At the attractive point there is a null vector and nearby that point the magnitude decreases. The function (5) fulfills all of these requirements and is used to calculate the vector magnitude in the vector field for a long-range attractive point.

$$f(x, y) = e^{-\frac{1}{\sqrt{x^2 + y^2}}} \tag{5}$$

The last attractive vector field is the short-range attractive point, which can be explained as a classical electrical force. The shorter the distance to the attractive point, the higher is the magnitude of the vector. Equation (6) presents the calculation of this vector field.

$$\mathbf{f}(x,y) = \frac{1}{x^2 + y^2} \cdot \begin{pmatrix} -x \\ -y \end{pmatrix} \tag{6}$$

These attractive vector fields are illustrated in Figs. 1, 2, 3, 4, 5 and 6. There are no axes with a scale in these figures, because they show the general vector fields which can be changed in position and size but the form stays always the same.

To realize obstacle avoidance there are two repulsive vector fields, a repulsive line and a repulsive point. The repulsive point vector field is calculated with function (7); at the repulsive point the magnitude tends towards infinity and at a greater distance the magnitude decreases to zero.

$$\mathbf{f}(x,y) = \frac{1}{x^2 + y^2} \cdot \begin{pmatrix} x \\ y \end{pmatrix} \tag{7}$$

The equation for the vector field of the repulsive line is similar to the repulsive point vector field and is shown in Eq. (8). The x-axis is defined to be the repulsive line, so just the distance towards the x-axis has influence on the magnitude of the vector.

$$\mathbf{f}(x,y) = \frac{1}{x \cdot \sqrt{x^2 + y^2}} \cdot \begin{pmatrix} x \\ 0.0 \end{pmatrix} \tag{8}$$

The magnitude of the repulsive vector fields guarantees collision avoidance, because near the point/line the repulsive vector dominates. These two repulsive vector fields are shown in Figs. 7 and 8.

2.2 Operations on Vector Fields

In order to build a complex vector field the atomic vector fields can be combined using the following set of operations:

- Addition
- Rotation
- Scalar
- Shift along x-axis
- Shift along y-axis
- Projection

Two vector fields can be added, for this simply the two vectors of both vector fields for one point are added. Next, a vector field can be rotated. The line following

S. Schmitt et al.

Fig. 7 Repulsive line

Fig. 8 Repulsive point

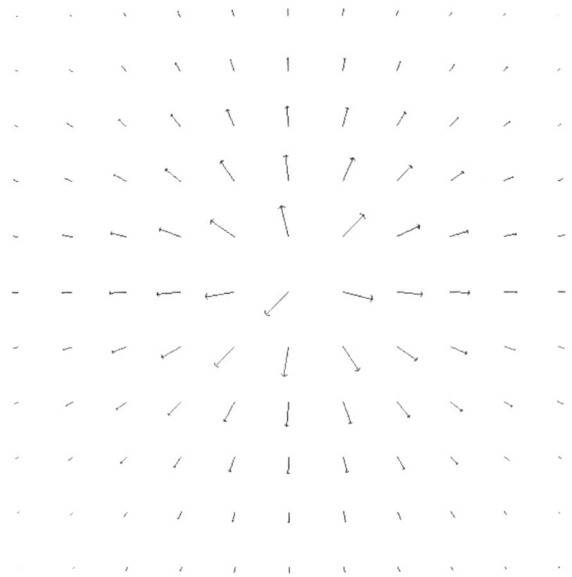

is not just possible in the default direction. The vector field for a line following can be rotated to have a line towards any direction. The scalar method is an enlargement of the vector field by a certain factor. With this method e.g. the attractive circle vector field (see Fig. 1) can be scaled up to have a bigger radius or the vector field for the repulsive point could be enlarged to have a bigger influence radius. The latter could be necessary if the obstacle to avoid is quite large.

The methods to shift along the x-axis and y-axis are used to position the vector field at the right point; a repulsive point should always be shifted towards the position of the corresponding obstacle.

Projection is an important method for a sailing boat robot. Sailing boats cannot go in every direction directly; there is a no-go zone around the direction where the wind is coming from. Vectors that point into this no-go zone need to be projected towards the next possible direction (generally $\pm45°$ towards the direction into the wind); the vector field just has vectors pointing towards sailable directions. Plumet et al. [8] also presents a potential field method, where the no-go zone of the wind is treated as a virtual obstacle. This repulsive field prevents the sailboat from going into the no-go zone. Yet it makes close hauled course manoeuvres unlikely. The projection method used in this approach is better for still being able to sail close hauled.

3   Control Architecture

The vector field needs to be constantly evaluated in order to regulate the marine robot direction. The robot embedded system above the plain vector field (low-level control) receives the vector after handing the current position of the boat to the vector field. It sets the actuators in order to steer to this direction. For sailboats this low-level controller corresponds to the controller described in [7]. High-level control needs to check conditions to be able to change to a next vector field respectively modify the vector field, when a waypoint is reached. This controlling strategy is illustrated in Fig. 9.

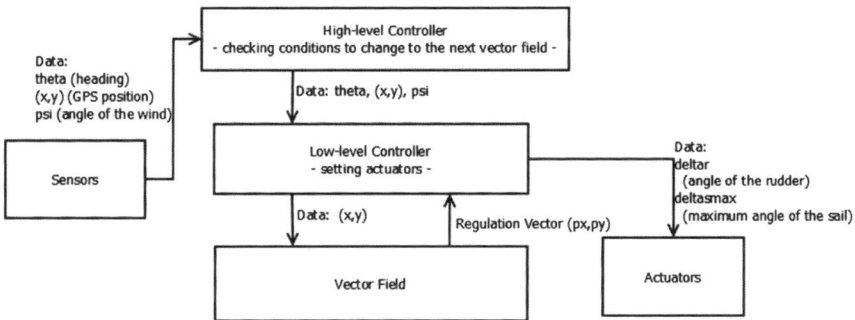

Fig. 9 Control architecture

This control loop consists of four main steps. In *step 1* the sensors of the boat
robot transmit the current values (e.g. GPS position and heading of the boat). This
is the left block in Fig. 9. The high-level controller checks conditions in a *step 2* and
potentially modifies the control vector field. *Step 3* is the computation of the vector
field. The input is the position (x, y) of the boat robot. The corresponding vector
is calculated, so that we get the desired direction as the output of this computation.
In *step 4* the low-level controller sets the maximum sail angle and the rudder angle
proportionally to the heading error (for the sailboat robot) respectively the motor
thrust and angle (for the motorboat). This is the setting of the actuators of the boat
(see arrow to the block on the right side of Fig. 9).

4 Experimental Validation

This control method has been simulated, subsequently implemented and tested at the
World Robotic Sailing Championship and International Robotic Sailing Conference
2013 in Brest, France. The algorithm has been successfully applied on the motorboat
of the team ENSTA Bretagne—Ifremer (Fig. 10).

In a first test the motorboat was regulated with a simple circle following vector
field. The path is shown in Fig. 11. After one circle the boat was brought of its trajectory (due to an encounter with another boat). It can be seen that it subsequently
returned to the circle line again.

One of the tasks, which the motorboat performed successfully, is the "Mobile
obstacle of known position avoidance task". In this task the motorboat robot has to
stay in a square DEFG (see Fig. 11) of 200 m. When an obstacle of known position (data transmission via XBee or Wifi) enters the square, the robot has to avoid a
collision and leave the square [9].

This task has been implemented using two vector fields; the first one is built with
repulsive lines as the borders of the square and the obstacle is added as a repulsive
point. The second vector field is a line following with the other obstacle as a repulsive

Fig. 10 Motorboat team
ENSTA Bretagne—Ifremer

Fig. 11 Path circle test

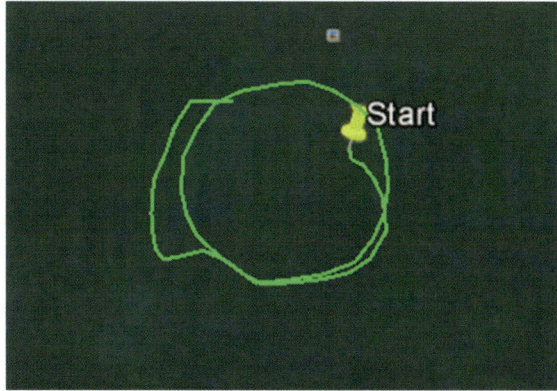

point. When the obstacle (which was a zodiac boat) entered the square the controller changed from the first to the second vector field, so that the robot there left the square. The angle of the line (rotation) has been calculated to not cross the path of the entering obstacle. The trajectory of the robot and of the zodiac (obstacle) are shown in Fig. 12. A black circle (number 1) shows the starting position. A short time before this point the program was started but the robot was still on the transporting boat. From this starting position the robot went to the middle of the square DEFG. The path of the zodiac is shown in red. When the zodiac enters the square between point D and G, the vector field of the motorboat robot changes to the line following (with the zodiac as a repulsive point) and the motorboat robot starts to leave the square. After the robot is 100 m away from the borders of the square, the robot automatically stops the motor, so that it is easier to pick up the motorboat robot. This completes the task.

Fig. 12 Path obstacle avoidance task

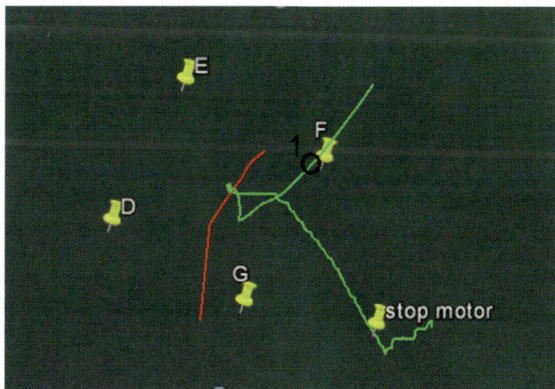

5 Discussion

A simple and fast obstacle avoidance control method based on vector fields has been presented.

There are a few disadvantages of potential fields like local minima and oscillation in narrow corridors [10] that are known from experience on land-based robots. Yet these problems are unlikely to occur at an ocean environment.

An advantage is that vector fields as a control allow the robot to take the best way towards the desired path no matter where the position is (there can be calculated a vector for each point). The arbitrary combination of vector fields gives the possibility to build the best suitable control vector field for the particular mission. Furthermore, it is easier to combine vector fields in order to avoid obstacles (so that we have one vector field for control) than to calculate multiple waypoints.

The algorithm can be also applied to sailboats. This was done in the simulation. Figure 13 shows a screen shot of a sailboat simulation. The right side shows the simulated sailboats floating on the water. On the left side there is a top view of the control vector field (of one of these sailboats) with the two boats presented by triangles.

The paper presents live tests conducted at the WRSC 2013 which has shown that the algorithm successfully implements obstacle avoidance. Compared to planning-based approaches, the vector field approach does consider the exact instantaneous position of obstacles. Moreover control parameters for the low-level control are calculated live and therefore the storage space is less than with a planned-based approach.

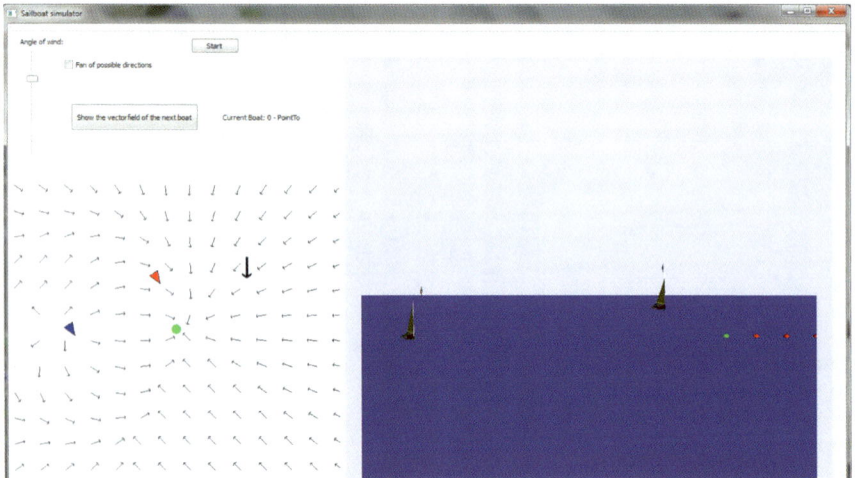

Fig. 13 Sailboat simulation

The simulation has been carried out with sailboat robots. In future this obstacle control algorithm should also be tested on a real sailboat robot. An enhancement of this algorithm could be to take the speed of the marine robot into account, achieving a faster reaction to a change in direction.

Acknowledgments I would like to thank Patrick Rousseaux and Olivier Ménage who helped me at the WRSC 2013 with the experimental part of my work. Helping at problems directly related to the boat they made it possible to try my algorithm successfully on the motor boat.

References

1. Bars, F. L., & Jaulin, L. (2012). An experimental validation of a robust controller with the VAIMOS autonomous sailboat. In *published in 5th International Robotic Sailing Conference* (pp. 74–84). Springer, Cardiff, Wales, England.
2. Cruz, N. A., & Alves, J. C. (2008). Ocean sampling and surveillance using autonomous sailboats. In *International Robotic Sailing Conference*, Austria.
3. Ménage, O., Bethencourt, A., Rousseaux, P., & Prigent, S. (2013). VAIMOS: realisation of an autonomous robotic sailboat. In F. Le Bars & L. Jaulin (Eds.), *Robotic sailing 2013* (pp. 25–36). Heidelberg: Springer.
4. Arkin, R. C. (1988). *Behavior-Based Robotics*.
5. Weissstein, E. W. (2013). Vector Field from MathWorld—A Wolfram Web Resource. http://mathworld.wolfram.com/VectorField.html.
6. Jaulin, L. (2012). Commande par espace d'état. *ENSTA Bretagne*, November 2012.
7. Jaulin, L., & Bars, F. L. (2012). A simple controller for line following of sailboats. In *Published in 5th International Robotic Sailing Conference* (pp. 107–119). Springer, Cardiff, Wales, England.
8. Plumet, F., Saoud, H., & Hua, M.-D. (2013). Line following for an autonomous sailboat using potential fields method. In *Published in IEEE Xplore*.
9. Bars, F. L. (2013). WRSC 2013 rules, 2013. http://www.ensta-bretagne.fr/lebars/wrsc2013/Rules_2013-08-28.pdf.
10. Koren, Y., & Borenstein, J. (1991). Potential Field Methods and Their Inherent Limitations for Mobile Robot Navigation. In *Published in Proceedings of the IEEE Conference on Robotics and Automation* (pp.1398–1404). Sacramento, California.

Printed by Printforce, the Netherlands